The Microbiomes of Humans, Animals, Plants, and the Environment

This series covers microbiome topics from all natural habitats. Microbiome research is a vibrant field of science that offers a new perspective on Microbiology with a more comprehensive view on different microorganisms (microbiota) living and working together as a community (microbiome). Even though microbial communities in the environment have long been examined, this scientific movement also follows the increasing interest in microbiomes from humans, animals and plants. First and foremost, microbiome research tries to unravel how individual species within the community influence and communicate with each other. Additionally, scientists explore the delicate relationship between a microbiome and its habitat, as small changes in either, can have a profound impact on the other. With individual research volumes, this series reflects the vast diversity of Microbiomes and highlights the impact of this field in Microbiology.

M. Pilar Francino
José Ángel Rufián Henares
Editors

The Microbiome and Personalized Nutrition

Springer

Editors
M. Pilar Francino
Department of Genomics and Health
Fisabio Foundation
Valencia, Spain

José Ángel Rufián Henares
Department of Nutrition and Food Science
University of Granada
Granada, Andalucía, Spain

ISSN 2662-611X ISSN 2662-6128 (electronic)
The Microbiomes of Humans, Animals, Plants, and the Environment
ISBN 978-3-031-93901-3 ISBN 978-3-031-93902-0 (eBook)
https://doi.org/10.1007/978-3-031-93902-0

This Springer imprint is published by the registered company Springer Nature Switzerland AG
The registered company address is: Gewerbestrasse 11, 6330 Cham, Switzerland

Preface

This book addresses researchers from academia and industry interested in the relationship between the gut microbiome, nutrition, and health. It offers an ample overview of the state-of-the-art in the study of the impact of the human microbiome in nutrition and how it can be leveraged in personalized diets. Its eight chapters cover the most relevant areas in this burgeoning field of inquiry, ranging from the manner in which foods and food supplements can be used to modulate the gut microbiome to the role of large-scale food information initiatives in bringing awareness of this topic to the general public. Special emphasis is given to the impact of food processing and cooking methodologies on gut microbiota modulation, an understudied area that will likely be crucial for developing sophisticated programs of personalized nutrition. In addition, we also address important societal aims, such as that of recycling agricultural products for the food industry with the specific goal of targeting the gut microbiome for better health. The role of microbiome-aware nutrition in the context of important diseases of the twenty-first century is explored in chapters dealing with obesity and allergy, especially in the case of children. The book also explores the potential contributions of metabolic modelling to guide dietary interventions for microbiota modulation, as well as how all this information can be incorporated into dietary applications that can reach the consumers interested in ameliorating their microbiome and their health.

Valencia, Spain — M. Pilar Francino
Granada, Andalucía, Spain — José Ángel Rufián Henares

Contents

Modulating the Gut Microbiome with Foods and Food Supplements

1

M. Pilar Francino

Abstract

In this chapter, we explore how the rapidly burgeoning scientific understanding of the gut microbiome can be applied in order to develop dietary strategies that optimize the composition and function of this bacterial community for better health and well-being. We review the many factors in the Western lifestyle that affect the integrity and ecological balance of the gut microbiome, and how dietary approaches can be used as a basic strategy to counteract their negative effects. To take maximum advantage of this strategy, it is important to understand the precise chemical composition of the foods we consume, in order to better understand how food components interact with the gut microbiome. This should facilitate the rational use of specific foods and food supplements to shape the microbiome in well-defined directions. In addition, food can also be supplemented with live microorganisms that fulfill or enable important functions within the gut microbiome. Overall, these dietary approaches have great potential for promoting well-being and for contributing to the prevention or treatment of the many diseases now known to be related to gut microbiome imbalances.

Keywords

Gut microbiome · Diet · Food supplements · Prebiotics · Probiotics

M. P. Francino (✉)
Department of Genomics and Health, FISABIO Foundation, Valencia, Spain

CIBER Epidemiología y Salud Pública, Instituto de Salud Carlos III, Madrid, Spain
e-mail: Pilar.francino@fisabio.es

M. P. Francino, J. Á. Rufián Henares (eds.), *The Microbiome and Personalized Nutrition*, The Microbiomes of Humans, Animals, Plants, and the Environment 6,
https://doi.org/10.1007/978-3-031-93902-0_1

All organisms live in close association with bacterial communities. In humans, the gut microbiome is the most complex microbiome in the body, achieving the highest cell densities recorded for any ecosystem (10^{11}–10^{12} cells/g of luminal content in the colon). It contains hundreds of different bacterial species, as well as archaea, eukaryotic microbes, and viruses. This complex microbial community participates in many crucial physiological processes in the human body and performs numerous beneficial functions related to nutrition and metabolism, protection against pathogens, gut development, and immune system modulation, among others (Sekirov et al. 2010; Clemente et al. 2012). These functions involve a variety of different mechanisms, which rely on the large repertoire of molecules produced by the microbiota that can affect gut function, gut communication with other organs, and many other high-level processes (Francino 2017; de Vos et al. 2022). Given its physiological importance, disturbances of the gut microbiome can initiate or compound the development of a variety of health disorders. For this reason, there is great interest in developing novel preventive and therapeutic strategies targeted toward modulating the gut microbiome for improved health and well-being. In light of the essential role of diet in shaping the gut microbiome, dietary approaches should constitute a basic strategy in this regard, including the rational use of specific foods and food supplements that promote the growth of beneficial bacteria and the optimal overall functioning of the gut microbial community.

1.1 Factors Affecting the Gut Microbiome

Importantly, the gut microbiome displays large variability among individuals, depending, among others, on age, diet, and health (Human Microbiome Project Consortium 2012). Microbiome composition and function are shaped by a combination of factors, both intrinsic (e.g., host genetics, immunity, and metabolic regulation) and extrinsic (e.g., lifestyle, medication, and diet). However, environmental factors tend to have a predominant effect over host genetics. In particular, diet is the environmental factor that produces the greatest effect on gut microbiota composition, explaining over 20% of the microbiota's structural variation (Rothschild et al. 2018). This highlights the potential for gut microbiota modulation through diet for disease management and the optimization of health and well-being.

The modern Western lifestyle affects the gut microbiota in numerous ways, including poor diet and generally decreased microbial exposure, which have resulted in a rapid loss of microbiota diversity (Vangay et al. 2018). The role of diet in shaping the overall diversity of the gut microbiota has been demonstrated (Smits et al. 2017; Sonnenburg and Backhed 2016; Sonnenburg et al. 2016), as well as its compounding effect over multiple generations (Sonnenburg et al. 2016). The diet of Western and Westernized populations has shifted drastically due to industrialization and the rise of large-scale food processing. In the past century, economic pressures for greater food production to support a growing world population have also resulted in agricultural practice changes that have generated a radical loss of dietary diversity (Heiman and Greenway 2016). The number of edible plant varieties and animal breeds being reared for human consumption has declined alarmingly, as farmers worldwide abandon their traditional choices and adopt genetically uniform, high-yielding counterparts to maximize production (Food and Agricultural Organization, United Nations, 2004 http://www.fao.org/3/a-y5609e.pdf). Moreover, exposure to pesticides employed in food crops and to antibiotics used as growth promoters for poultry, swine, and cattle is likely to further affect the human gut microbiome. These alterations of agricultural practices and the advent of industrial food processing have resulted in a Western diet characterized by a steep decrease in dietary diversity, reduced fiber consumption, and high intakes of sodium, fat, and simple sugars. This diet has been strongly associated with alterations in gut microbiota composition and with related pathologies such as obesity and metabolic diseases, although the biological mechanisms are still unclear (Zinöcker and Lindseth 2018). Health problems associated with the Western diet started as a concern of developed societies, but they are currently also prevalent in developing countries, where economic constraints will likely worsen their impact.

Modern Western practices related to infant birth and rearing, especially C-sections and formula feeding, also have notorious effects on the development of the gut microbiome. These early microbiome alterations have been associated with defects in immune, metabolic, and neurologic development and related health outcomes, such as atopic diseases, obesity, and autism. Therefore, it will be crucial to remember that in all likelihood, the best opportunities for successful health promotion and disease prevention through microbiome modulation will be very early on, during infancy and even prenatally (Moriki et al. 2022). Medical drug treatments are another important extrinsic factor that can negatively alter the microbiome. The most representative example of this is antibiotic use, which is known to cause short-term disturbances in the microbiome and to have a long-term impact that may last for up to 2 years (Jernberg et al. 2007). Notably, when antibiotic use induces a severe alteration in the composition and function of the microbiome (i.e., dysbiosis), it can increase the risk of developing a variety of associated health problems. Numerous studies have documented that antibiotic-induced dysbiosis can promote an increased susceptibility to infections (e.g., by *Clostridioides difficile*) and immune and metabolic disorders (Francino 2016). In particular, exposure to antibiotics in early life is a risk factor for the development of inflammatory bowel disease (IBD) (Hviid et al. 2011) and obesity (Trasande et al. 2012; Bailey et al. 2014; Saari et al. 2015), indicating that the associated loss of microbial diversity or the lack of some specific microbiota organisms has long-lasting health consequences.

Overall, the microbiome alterations associated to the modern Western lifestyle are thought to be linked to the generalized rise in immune and inflammatory diseases that Western countries have undergone in recent decades. As cases in point, the prevalence of inflammatory bowel diseases (IBD), obesity, and type 2 diabetes has sharply increased during the past 60 years, and for each of these diseases a reduction in gut microbiome diversity has been documented (Ott et al. 2004; Turnbaugh et al. 2009; Larsen et al. 2010). Other rising non-communicable diseases (NCDs) that have been linked to microbiome alterations include psychological disorders, auto-immune diseases, cardiovascular diseases (CVD), and colon cancer.

1.2 Food Composition and the Gut Microbiota: Macronutrients, Micronutrients, and Phytochemicals

In addition to providing the necessary nutrients to the host, foods consumed in the diet also serve as a substrate for the microbial community of the gut. The main macronutrients used as energy sources in the human diet are carbohydrates, proteins, and fats. The gut microbiota has the enzymatic capacity to degrade carbohydrates resistant to host enzyme digestion, as well as proteins, lipids, and other molecules such as polyphenols, scavenging between 10% and 30% of the ingested energy (Adak and Khan 2019). However, the digestibility and accessibility of these diverse compounds to the microbiota can vary considerably. The main substrates for the gut microbiota are the nondigestible carbohydrates that constitute dietary fiber. These reach the colon partially degraded or undigested, where they will be partially or totally fermented by the microbiota. Saccharolytic bacterial fermentation produces metabolites that are generally beneficial for the host. In contrast, if dietary carbohydrates are scarce, bacteria turn to alternative energy sources and produce potentially detrimental metabolites. In particular, microbial protein metabolism results in some fermentation products that can be harmful to host health. Colonic bacteria differ in their degree of metabolic specialization, with some able to metabolize a large range of substrates, while others specialize in activities such as the primary degradation of plant cell walls or the utilization of carbohydrates and peptides from the intestinal mucus layer. Other bacteria specialize in cross-feeding on the products of complex carbohydrate degradation or fermentation by other colonic species (Scott et al. 2013).

The main products resulting from bacterial fermentation of dietary carbohydrates in the colon are gases and short-chain fatty acids (SCFAs). The production of these weak acids tends to lower the colonic pH, influencing microbiota composition. The main SCFAs produced are acetate, propionate, and butyrate in ratios ranging from 3:1:1 to 10:2:1 (Rowland et al. 2018). All three SCFAs display very important, but distinct, functions in the body. Many bacteria of the microbiota produce acetate, making it the most abundant SCFA in the gut. Acetate is also essential for the growth of other bacteria, making it an important metabolite for maintaining cross-feeding networks. In the host, acetate is transported to peripheral tissues and used in cholesterol metabolism and lipogenesis, and studies in mice indicate that it also contributes to the central regulation of appetite (Frost et al. 2014). In contrast to acetate, propionate and butyrate tend to be produced by more limited numbers of specific bacteria, which usually have the capability to produce one or the other, but rarely both (Reichardt et al. 2014). Remarkably, propionate and butyrate can result not only from fermentation of carbohydrates but also from fermentation of peptides and amino acids from dietary proteins (Louis and Flint 2017). The major producers of propionate in the gut are members of the two main bacterial phyla of the microbiota, Bacillota and Bacteroidota (previously referred to as Firmicutes and Bacteroidetes), including members of the Negativicutes class and of the genera *Clostridium* and *Bacteroides*. Propionate serves locally as an energy source for epithelial cells but is also used in gluconeogenesis, either in the intestine or after transport to the liver (De Vadder et al. 2014). Also, like acetate, propionate is likely involved in satiety signaling (Brown et al. 2003; Tazoe et al. 2008; De Vadder et al. 2014). On the other hand, the major butyrate producers of the gut microbiota are Bacillota members, mainly of the family Lachnospiraceae and the Oscillospiraceae genus *Faecalibacterium prausnitzii*, although metagenomic analyses have identified many other species containing butyrate production pathways. Butyrate may have the most critical roles for human health of all SCFAs, as it represents the main source of energy for human colonocytes, activates intestinal gluconeogenesis with beneficial effects on glucose and energy homeostasis (De Vadder et al. 2014), and has potential anticancer activity through induction of apoptosis of colon cancer cells (Steliou et al. 2012), among many other important functions.

Altering the carbohydrate content of the diet can affect the gut microbiota composition and the profile of SCFAs it produces. Reducing dietary intake of carbohydrates results in decreased abundances of butyrate and butyrate-producing bacteria of the Lachnospiraceae family, such as *Roseburia* and *Agathobacter rectalis* (previously referred to as *Eubacterium rectale*) (Duncan et al. 2007). In contrast, dietary supplementation with wheat bran, which consists mainly of arabinoxylan oligosaccharides, increases the abundance of all three main SCFAs (Francois et al. 2012). Substantial changes in gut microbiome composition have also been measured in response to the addition of specific individual carbohydrates to the diet. For instance, the increase of resistant starch modifies the composition of the microbiota, metaproteome, and metabolome. In terms of taxonomic composition, resistant starch favors an overall increase of the Bacillota over the Bacteroidota, including increases in the relative abundances of specific members of the Bacillota that have widely evidenced beneficial effects on health, such as *Faecalibacterium prausnitzii*. Moreover, resistant starch impacts lipid metabolic pathways carried out by both the host and the microbiome, including those for fatty acid metabolism, primary and secondary bile acid biosynthesis and secretion, steroid biosynthesis, and metabolism of linoleic and arachidonic acid. This indicates that the direct products of the colonic fermentation of resistant starch are further metabolized by the gut microbiome and involved in lipid biosynthesis and in the metabolism of host-derived lipids (Maier et al. 2017).

Micronutrients such as vitamins and minerals commonly found in the diet may also reach the microbiota and be used by some of its members to grow, thereby affecting the composition and functionality of the intestinal microbial community. For instance, supplementation with vitamins B, C, D, and E changes the composition of the microbiota by promoting the growth of health-associated bacteria such as *Bifidobacterium*, *Lactobacillus*, and *Roseburia*. Similarly, minerals such as calcium, iron, zinc, magnesium, and phosphorus may have the potential to influence the microbiota (Lv et al. 2016; Bielik and Kolisek 2021; Lagoumintzis and Patrinos 2023). On the other hand, the gut microbiota is involved in the synthesis of several vitamins, such as biotin, folate, vitamin K, and some B vitamins, which are necessary for many physiological processes (Rowland et al. 2018). The gut microbiota also influences the absorption of essential minerals such as calcium, iron, and magnesium by increasing their bioavailability (Barone et al. 2022).

Besides macro- and micronutrients, foods naturally contain hundreds of additional chemical compounds, and processed foods will also include many other nonnutritive components such as preservatives and other food additives. Plant-based foods contain a large variety of phytochemicals, bioactive compounds produced by plants for their protection through their primary or secondary metabolism. Thousands of different phytochemicals are known (Rutz et al. 2022), from sources such as whole grains, fruits, vegetables, nuts, and herbs. The possible health effects of most of them have not yet been established, although many are being investigated. Phytochemicals can be classified into various major categories, including carotenoids and polyphenols. In particular, vegetables and fruits contain a large variety of polyphenols, including phenolic acids, flavonoids, and lignans. These polyphenols exhibit great structural diversity, which impacts their bioavailability, metabolism, and bioactivity (Manach et al. 2004). Some polyphenols have been described to have antioxidant capacity and antimicrobial activity. Also, evidence from animal models and in vitro experiments shows that many polyphenols are extensively metabolized by the colonic microbiota, which can affect their bioactivities (Russell et al. 2008; Duda-Chodak et al. 2015). Most polyphenols, especially flavonoids, are present in food as glycosides (i.e., conjugated to various types of sugars), although some can be esterified with organic acids or lipids or form high molecular weight oligomers and polymers. These conjugated and polymeric forms are generally poorly bioavailable and are not absorbed by the host. Rather, they pass into the colon where they are metabolized by consortia of microbes. First, they are hydrolyzed by species of *Bacteroides* (*B. distasonis*, *B. uniformis*, *B. ovatus*) and *Eubacterium* (*E. cellulosolvens* and *E. ramulus*), among others (Braune et al. 2015; Marín et al. 2015). Hydrolysis products can then be further metabolized by other components of the colonic microbiota via dehydroxylation, decarboxylation, and ring breakage to generate simpler phenolic compounds, such as hydroxyphenyl-acetic acids and hydroxyphenylpropionic acids (Rowland et al. 2018). Therefore, dietary polyphenols have the potential to play an important role in modulating the gut microbiota, as they can inhibit the growth of specific taxa through their antimicrobial activities while enhancing the growth of others that can use them or their metabolic products as substrates.

1.3 Impact of Broad Dietary Patterns

Cross-sectional studies have revealed consistent microbiome differences when comparing populations with widely divergent diets and lifestyles, such as Westernized and non-Westernized populations. In particular, a higher prevalence of the Bacteroidota genus *Prevotella* has been consistently detected in non-Westernized populations (De Filippo et al. 2010; Yatsunenko et al. 2012; Schnorr et al. 2014; Obregon-Tito et al. 2015; Smits et al. 2017; Vangay et al. 2018; Hansen et al. 2019; Pasolli et al. 2019; Tett et al. 2019). These populations typically consume diets rich in fresh foods, including a large proportion of vegetables and fruits, with a high content in fiber and complex carbohydrates. In contrast to the enrichment of

Prevotella in non-Westernized populations, as well as in individuals in Westernized areas following a vegetable-based diet (De Filippis et al. 2016), individuals consuming a typical Western diet are commonly enriched in the related genus *Bacteroides*. However, this paradigm may be an oversimplification, as studies have produced conflicting results (Herman et al. 2019) and detailed taxonomic analyses suggest that different species of *Prevotella* and *Bacteroides* can have different dietary association patterns, including links to both plant-based and animal-based diets within each genus (De Filippis et al. 2016). Nevertheless, metagenomic studies capable of species-level resolution have shown that in Westernized populations the *Prevotella* species *P. copri* is indeed less prevalent (Pasolli et al. 2019; Vangay et al. 2018) and has underrepresented intraspecific diversity (Tett et al. 2019). *P. copri* comprises four distinct subspecific clades with different carbohydrate metabolism repertoires that confer a large overall diversity in plant-derived carbohydrate utilization. Moreover, it has been shown that *P. copri* provides a benefit to its host in response to a high-fiber diet (Kovatcheva-Datchary et al. 2015; De Vadder et al. 2016), but not to one high in fat (Pedersen et al. 2016). Therefore, it seems likely that the fiber reduction and fat increase in the Western diet have resulted in an inability to sustain the original prevalence and diversity of this species (Tett et al. 2019).

Long-term diet-microbiota links have also been widely analyzed in cross-sectional epidemiological studies within single populations through the use of food records and food frequency questionnaires. These instruments allow evaluation of the effect of diet on the gut microbiota through different analytical approaches involving the definition of broad dietary patterns and dietary indices, or the analysis of gut microbiota associations with nutrients, foods, and food groups. Such analyses have corroborated that long-term dietary patterns are strongly associated with gut microbiota composition, with carbohydrate consumption linked to *Prevotella* and protein and animal fat linked to *Bacteroides* (Wu et al. 2011). In addition to cross-sectional studies, diet-microbiota associations have also been established in longitudinal studies that have followed how microbiota composition and food choices covary temporally within subjects. Johnson et al. (2019) showed that an individual's dietary pattern could inform the prediction of their microbiota composition the following day, and that, as dietary intake in most subjects varied daily, this prediction was improved by considering at least the 2 days prior. Ongoing research aimed at further understanding microbiota changes in response to habitual diets will be fundamental in order to improve dietary strategies for sustained microbiota shifts.

A notable dietary pattern associated with a variety of health benefits that may partially result from a beneficial modulation of the gut microbiome is the Mediterranean diet (MedDiet) (De Filippis et al. 2016; Mitsou et al. 2017). The MedDiet is a healthy dietary pattern that has been associated with a reduced risk of numerous NCDs (Sofi et al. 2008). It is based on the food patterns followed by Mediterranean populations in the early 1960s, involving high consumption of fruits, vegetables, legumes, unrefined cereals and nuts, moderate consumption of fish, poultry and dairy products (mainly cheese and yogurt), low consumption of red meat products, use of olive oil as the main edible-fat source, and moderate but regular consumption of wine (Willett et al. 1995). These characteristics provide a large content and variety of constituents, such as carbohydrates, fiber, unsaturated lipids, and antioxidants, likely to have a beneficial impact on gut microbiota composition and function, positively impacting overall health. An example in point is the link between the MedDiet and a reduced risk of allergic diseases in children, including food allergies, asthma and, atopy. This is probably due to the high content of fiber contained in fruits, cereals, and legumes, which, upon fermentation by intestinal bacteria, results in butyrate and other SCFAs that have numerous immunoregulatory properties. On the contrary, the Western diet has been linked with a higher incidence of allergic diseases, likely as a result of reduced levels of butyrate (Rey-Mariño and Francino 2022). Adherence to the MedDiet can be measured through composite indices aimed at capturing the consumption of a variety of food items positively or negatively associated with this dietary pattern, such as the MedDietScore (Panagiotakos et al. 2006). Thanks to these indices, various studies have indeed linked MedDiet adherence to increased abundance of *Prevotella* and fecal SCFAs (De Filippis et al. 2016; Gutiérrez-Díaz et al. 2016; Shankar et al. 2017) and to decreased abundance of *Escherichia coli*, a typical pathobiont or opportunistic pathogen of the gut (Mitsou et al. 2017). Analyses of other long-term polysaccharide-rich diets have also been linked to an underrepresentation of *Escherichia* and other Enterobacteriaceae (De Filippo et al. 2010), and dietary antioxidants may inhibit the growth of these bacteria (Duda-Chodak et al. 2015). However, in a systematic literature review of the randomized controlled trials (RCTs) and observational studies that have explored how the MedDiet impacts the gut microbiota, consistent effects across studies could not be clearly established (Kimble et al. 2023). This may relate to variation in the populations analyzed, the adherence scores used, and the overall methodology and quality of the different studies.

In addition to scores that measure adherence to specific dietary patterns, overall dietary diversity indices are also among the important tools employed to study diet–microbiota interactions. High dietary diversity has been associated with a diverse (Claesson et al. 2012) or stable (Johnson et al. 2019) gut microbiome—properties thought to be health-promoting (Heiman and Greenway 2016). Indeed, the diversity of the microbiota, which describes its species richness and complexity, is generally associated with host health (Chatelier et al. 2013) and it likely provides greater resilience for the ecosystem to recover from or adjust to perturbations. Microbiota diversity has also been directly linked with the diversity of the bile acids it produces, and therefore likely contributes to regulating the available pool of these molecules (Jones et al. 2008; Sayin et al. 2013). The number of different types of plant-based foods consumed has also been associated with higher microbiota diversity and a larger presence of the SCFA-producing bacteria such as *Faecalibacterium prausnitzii* and *Oscillospira* (McDonald et al. 2018; Johnson et al. 2019). In contrast to diversity, measures of dietary stability (i.e., to what degree food intake changes from day to day) have not been found to correlate with microbiome stability. Even in the extreme case of subjects consuming meal replacement shakes of defined composition, their day-to-day microbiome changes were not seen to differ from those of individuals consuming regular diets (Johnson et al. 2019).

Nutrient profiles derived from food questionnaires are also used to study the effects of diet on the gut microbiota. This approach supports fiber as the nutrient most widely accepted to produce a beneficial effect on microbiota composition. Individuals with high-fiber diets have higher levels of fermentable substrates available, and dietary fiber has been widely described as an essential nutrient for maintaining the diversity of the microbiota (Flint et al. 2012). In contrast, a low-fiber diet deprives the gut microbiota of external energy sources and forces it to use endogenous fermentable substrates such as the mucus that lines the colon epithelium. The long-term elimination or drastic reduction of fiber from the diet results in microbial losses that are difficult to reverse (Hill et al. 2017; Reddel et al. 2019). Claesson et al. (2012) studied individuals following a low-fat/high-fiber diet and described that it was associated with improved health parameters and with the most diverse microbial profiles in comparison with individuals following moderate-fat/low-fiber or high-fat/low-fiber diets. Fiber source also has an influence, as fibers from different fruits and grains have been shown to have significantly different impacts on the gut microbiota (Johnson et al. 2019).

Unfortunately, diets that exclude essential nutrients for medical reasons—such as gluten-free diets for celiac disease or diets low in fermentable oligo-di-monosaccharides and polyols (FODMAPs) for irritable bowel syndrome (IBS)—may also decrease the diversity of the gut microbiota. For example, it has been reported that gluten-free diets decrease the abundance of health-associated bacteria, whereas detrimental bacteria increase due to the reduction in polysaccharide intake (Sanz 2010; Hill et al. 2017; Reddel et al. 2019).

1.4 Impact of Dietary Interventions

Complementing the studies based on analysis of habitual diet through food records, dietary interventions have also shed light on diet–microbiota relationships and will likely be an important strategy to initiate—if maybe not maintain—health-promoting microbiota shifts. Numerous interventions have been performed with the aim of assessing the impact of specific diets, foods, nutrients, or other food ingredients on the intestinal microbiota of the participants. Short-term dietary interventions involving severe dietary restriction have been shown to be capable of rapidly altering the human gut microbiota. However, the observed microbiome shifts are usually transient and do not persist beyond a few days, as seen in the study by David et al. (2014), where the effects of two dietary regimens, one almost exclusively plant-based and the other almost exclusively animal-based, were compared in a cross-over design for 5 days. The gut microbiota can also rapidly respond to altered macronutrient levels and novel food components, as Wu et al. (2011) showed by investigating the effect of high-fat/low-fiber and low-fat/high-fiber interventions, where the participants also displayed substantial but transient shifts.

The microbiota's resilience to perturbations induced by short-term dietary changes may depend on the responsiveness of certain key taxa. This resilience may protect against the loss of critical taxa, but it may also inhibit the improvement of an established dysbiotic microbiota. It is unclear how long a dietary intervention would need to last in order to overcome resilience and permanently shift the ecological homeostasis of the microbiota, creating a sustainable new state (Leeming et al. 2019). However, given the strong association of long-term dietary patterns with gut microbiota composition, it seems likely that dietary strategies aimed at sustained microbiota shifts that can affect health outcomes should require medium- to long-term interventions. For instance, interventions involving consumption of the MedDiet for periods of 2 months to 2 years have been successful in promoting positive gut microbiome changes in overweight or obese subjects (Haro et al. 2017; Meslier et al. 2020). Moreover, it has been observed that the function and composition of the gut microbiota may be affected by variation at the scale of yearly seasons, as occurs in hunter-gatherers (Smits et al. 2017).

Another important consideration is that individuals are highly heterogeneous with respect to their ability to respond to dietary interventions. One of the greatest challenges to understanding the relationship between diet, the microbiota, and health is deciphering the high variability in individual responses to dietary intake, leading to conflicting results in studies aimed at finding diet-specific effects (Healey et al. 2017; Leeming et al. 2019). This variability is likely linked to the high interindividual heterogeneity in baseline dietary habits and composition of the gut microbiota. For instance, the responsiveness of the gut microbiota to an increase in fermentable substrate may be influenced by an individual's previous dietary intake and resulting microbiota composition. Indeed, inulin supplementation results in a greater increase in bifidobacteria when their basal abundance is low (Tuohy et al. 2001; Healey et al. 2017). Moreover, blood glucose responses to foods also appear to be personalized and may depend on the individual's microbiome composition (Zeevi et al. 2015).

1.5 Effects of Specific Foods and Food Groups

Although it is well established that diet is a key determinant of human gut microbiome composition, the fine-scale effects of specific foods are not well understood. However, analyses of dietary records indicate that microbiome composition is more strongly associated with concrete food choices than with overall nutrient profiles (Johnson et al. 2019). Food choices are much more highly personalized and even variable over time within subjects than the associated nutrient profiles. Conventional profiles of macro- and micronutrients do not reflect important aspects of food composition that may have significant effects on gut microbes (Barratt et al. 2017), such as the abundance and variety of polyphenols and other nonnutritive chemical compounds. Moreover, plant and animal tissues consumed as foods may contain exosomal microRNAs that can affect microbial metabolism and composition in the gut (Teng et al. 2018). Even for common macronutrients, each general macronutrient type can be represented in different foods by a variety of specific molecules. Accordingly, it has been established that not only the amounts but also the types and diversity of carbohydrates that reach the colon affect the composition of the microbiome (Walker et al. 2011; Scott et al. 2013) and the metabolites it produces (Flint et al. 2012). Indeed, specific types of polysaccharides can have highly unique conformations and very different levels of accessibility to the microbiota. Moreover, although most gut bacteria are adapted to use glycans as their main energy source, different glycan sets are preferentially used by different microorganisms. As different foods provide different glycan combinations, they will select for specific gut bacteria. In the case of fiber, for example, this term regroups many different chemical structures that can be used by specific bacteria with differing metabolic capacities. Accordingly, dietary analyses indicate that individuals consuming similar specific fiber-rich foods have similar microbiota compositions (Johnson et al. 2019). Therefore, it is an oversimplification to assume that diets similar in overall nutrient composition will produce the same effects on the gut microbiota without taking into account the specific kinds of foods in the diet and the molecules they contain. On the contrary, specific food choices could be leveraged to fine-tune microbiome composition and function. Moreover, cooking and other types of food processing will also alter the structure and composition of foods, changing their digestibility and nutrient bioaccessibility (Chen et al. 2020; Navajas-Porras et al. 2020), with likely consequences for their effects on the gut microbiota (Pérez-Burillo et al. 2018; Lerma-Aguilera et al. 2024; Lerma-Aguilera 2024). Cooking and other food processing techniques may therefore provide yet another layer of choices for the rational design of diets aimed at modulating the gut microbiota. The next chapter in this book will consider extensively the various effects of food processing on gut microbiota composition and functionality.

However, the choice of foods in diets targeting gut microbiota modulation also faces the challenge of interindividual variability. Longitudinal studies in individual subjects have established that, like responses to dietary interventions, specific food–microbe interactions tend to be highly personalized (Johnson et al. 2019). In fact, this likely contributes to the heterogeneity of microbiota responses observed in interventions and reduces the detected effect sizes of diet in population-level studies. A high level of personalization in food–microbe interactions has also been evidenced in in vitro analyses. Lerma-Aguilera et al. (2024) employed an in vitro digestion–fermentation model to expose fecal microbiota from different healthy adult individuals to a variety of representative foods of the Mediterranean and Western diets, each one raw or cooked in several different ways. Their results showed that many food-induced microbiota changes were individual-specific, especially those that depended on the method of cooking. Moreover, similar work with samples from healthy children and children suffering from different pathologies associated with, or resulting from, dietary disturbances (obesity, food allergy, celiac disease) showed that the differences in microbiota composition among these children resulted in different responses to foods. Importantly, in these experiments, the microbiota of children changed more in response to different foods than that of adults, suggesting greater malleability through diet. This could be related to the lower diversity of children's samples, as less diverse communities are generally less resilient to perturbations (Lerma-Aguilera 2024).

Nevertheless, in spite of interindividual variability, it seems likely that at least some specific foods should tend to have consistent effects in terms of selecting for specific microbes or for microbes encoding similar functional capabilities. In agreement, the in vitro analyses mentioned above (Lerma-Aguilera et al. 2024; Lerma-Aguilera 2024) revealed several shared trends observed across samples from diverse individuals, which likely represented robust effects. Notably, plant-based and animal-based foods as a whole had clearly different impacts on the abundances of bacterial taxa, and differences were also associated with specific food categories and individual foods. Overall, fermentation of animal-based foods as well as vegetable fats caused the most significant alteration on the microbiota of both adults and children, probably due to their lack of carbohydrates. For instance, butter, fish, and cheese led to increases in many taxa within the Lachnospiraceae family, including beneficial bacteria but also the genus *Lachnoclostridium*, associated with several diseases. Moreover, longitudinal in vivo studies have shown that multiple bacterial species within the same taxonomic family can correlate with foods from the same food group, and that related foods may share properties—beyond simple nutrient composition—which allow them to promote bacteria with similar characteristics (Johnson et al. 2019). Biochemical analyses of food composition that go beyond conventional nutrient profiles—for instance, through mass spectrometry—could help understand similarities among foods that may inform prediction of their effects on the microbiota.

Interactions between different foods are also likely to be very important in terms of the effect of diet on the gut microbiota. For instance, foods such as shellfish, eggs, milk, red meat, and poultry are rich in phosphatidylcholine, which is converted by intestinal microbes to trimethylamine (TMA) and then oxidized by the host to the atherosclerosis-associated trimethylamine-N-oxide (TMAO) (Wang et al. 2011). However, this process can be reduced by including dietary foods common in the Mediterranean diet such as balsamic vinegar, red wine, cold-pressed extra virgin olive oil, or grapeseed oil, which contain an inhibitor of TMA production (Wang et al. 2015).

1.6 Use of Functional Foods and Food Supplements

Foods can naturally contain, or be enriched with, functional ingredients that provide physiological benefits beyond basic nutrition. Some of these benefits may result from modulating the composition and functionality of the gut microbiota. These functional ingredients can also be consumed as food supplements, that is, pills, tablets, capsules, liquids, or other concentrated formats designed to be taken in small, measured amounts. Food supplements are legally considered foods, the purpose of which is to supplement the normal diet.

1.6.1 Prebiotics and Other Microbiota-Targeted Functional Compounds

Prebiotics are potential components of functional foods and food supplements that can be employed to modulate the gut microbiota. A prebiotic is a substrate that is selectively utilized by host microorganisms, conferring a health benefit (Gibson et al. 2017). This broad consensus definition, agreed upon by a panel of experts in microbiology, nutrition, and clinical research, encompasses different types of substances, including noncarbohydrates. The important requirements for a substance to be considered a prebiotic are that beneficial health effects must be documented and shown to come about through selective microbiota-mediated mechanisms. Specifically, prebiotics should fulfill three criteria: (a) resistance to gastric acidity, hydrolysis by mammalian enzymes, and gastrointestinal absorption; (b) fermentation by intestinal microbiota; and (c) selective stimulation of the growth and/or activity of intestinal bacteria associated with health and well-being. Thus, prebiotics differ from most dietary fibers such as pectins and xylans, which are broadly metabolized and encourage growth of a wide variety of gut microorganisms, irrespective of their health benefits (Gibson et al. 2017). However, certain fermentable fibers could be considered prebiotics (Delcour et al. 2016), if they are shown to be selectively utilized by host microorganisms that promote health. Also, antibiotics, minerals, vitamins, and bacteriophages, which are not bacterial growth substrates, cannot be considered prebiotics, even though their intake might alter microbiota composition and function (Gibson et al. 2017).

Substances that can act as prebiotics are broadly found across plants but can also be produced by animals (such as the oligosaccharides present in milk) and yeasts. Prebiotics include mainly various types of carbohydrates that are resistant to degradation in the small intestine and are fermented by microbes in the colon into SCFAs, gases, and other products (Scott et al. 2013). Fructans (fructo-oligosaccharides [FOS] and inulin) and galactans (galacto-oligosaccharides [GOS]) were the first to be recognized for their ability to modulate the microbiota to the benefit of the host (Gibson and Roberfroid 1995), with their effects acting through the enrichment of *Lactobacillus* and/or *Bifidobacterium* spp. Numerous studies have investigated the prebiotic effects of FOS and GOS. In particular, two of the best-studied prebiotics are oligofructose and inulin, fructose polymers with varying degrees of polymerization that are widely used as plant storage carbohydrates. They have been shown to stimulate the growth of intestinal bifidobacteria and the production of SCFAs (Flamm et al. 2001) and to contribute to the regulation of food consumption and facilitate weight loss (Delmée et al. 2006).

More recently, high-throughput sequencing analyses have been able to show that prebiotics usually affect a broader range of beneficial microorganisms beyond lactobacilli and/or bifidobacteria. For instance, inulin-type fructans can result not only in the increase of bifidobacteria but also in that of other beneficial bacteria such as *F. prausnitzii* and *Anaerostipes* spp., and in the decrease of detrimental bacteria such as *Bilophila* spp. (Dewulf et al. 2013; Vandeputte et al. 2017). On the other hand, some prebiotics with increased specificity have been described that can favor colonization by better-defined groups of gut bacteria. For example, seaweed polysaccharides have been shown to enrich for specific *Bacteroides* spp., while certain GOS can enrich for specific strains of *Bifidobacterium adolescentis* (Wang et al. 2020).

The changes in microbial composition and metabolite production brought about in the gut microbiome by prebiotic administration can impact numerous host physiological processes and result in a variety of benefits, both local and systemic. In the gut, prebiotic-induced changes can contribute to inhibition of pathogens, immune stimulation, and improvements in bowel function. Beyond the gastrointestinal tract, documented benefits of prebiotics include cardiometabolic effects (reduction of blood lipid levels and insulin resistance), regulation of appetite and satiety, improvement of mineral bioavailability for bone health, and production of metabolites that improve brain function, among others (Gibson et al. 2017; Cunningham et al. 2021).

The prebiotic concept has recently expanded to potentially include new substances that can be employed for microbiota modulation and health promotion. Although most currently established prebiotics are carbohydrate-based, the concept may now encompass noncarbohydrate substances, including some polyphenols and other phytochemicals, which are under research for their prebiotic potential. Most polyphenols are not absorbed in the small intestine and reach the colon (Clifford 2004), where they are extensively metabolized by the microbiota. Although the interaction between polyphenols and the microbiota is bound to be complex and is only beginning to be understood, evidence is emerging on the capacity of dietary polyphenols to modulate microbiota composition and activity. For instance, polyphenol-rich cranberry extracts have been shown to stimulate the health-associated *Akkermansia muciniphila* (Anhê et al. 2015) and to provide antimicrobial action against pathogens (Kumar Singh et al. 2019). Also, we and others have investigated the effect of tannins, one type of polyphenols, on the gut microbiota through in vitro fermentations by combining them with different foods (Molino et al. 2021), or evaluating different methods of tannin release (Molino et al. 2023), as well as through in vivo studies involving short-term supplementation in healthy patients (Molino et al. 2022). These studies have shown that tannins are associated with beneficial effects on the gut microbiota through their antioxidant activity and the stimulation of SCFA production. In accordance with these and other studies, substantial evidence exists for an association of polyphenol intake with a reduced risk of chronic diseases (Cardona et al. 2013).

Importantly, increasing evidence indicates that the health benefits associated with polyphenol consumption depend on their selective metabolization by the gut microbiota and are often not directly provided by the ingested polyphenols but rather by their microbial metabolites (Duenas et al. 2015). For instance, equol is more bioactive than its parent polyphenol molecule, the soy isoflavone daidzein, in a range of areas including estrogen-related activity, antioxidant capacity, and potential anticancer effects (Landete et al. 2015). This dependence on bacterial metabolism reinforces the notion that polyphenols can be considered prebiotic compounds. However, more studies of the health benefits and mechanisms of action of these emerging prebiotics are required to confirm their prebiotic status (Gibson et al. 2017).

Maybe not surprisingly, as with other dietary interventions and in vitro fermentation studies, those involving dietary plant polyphenols—including flavonoids, isoflavones, lignans, hydroxycinnamic acids, ellagitannins, and anthocyanins—have revealed large interindividual variation in their absorption, metabolism, and excretion (Rowland et al. 2003; Gill et al. 2010; Tomas-Barberan et al. 2014; Quartieri et al. 2016). Variation in the colonic metabolism of polyphenols is likely related to differences in gut microbiota composition, which could therefore impact the potential health benefits of these phytochemicals. In the aforementioned case of the soy isoflavone daidzein, variation in the capacity of the microbiota to produce equol may define whether soy consumption does or does not provide benefits in terms of bone health, menopausal symptoms, and breast cancer (Lampe 2009).

1.6.2 Use of Probiotics in Foods and Food Supplements

The Food and Agriculture Organization and the World Health Organization define probiotics as "live microorganisms that, when administered in adequate amounts, confer a health benefit on the host" (http://www.who.int/foodsafety/publications/fs_management/en/probiotics.pdf) (Hill et al. 2014). This definition encompasses a wide range of microbes and health applications. The effects of probiotics can be very diverse, but every microbial species that is considered a probiotic must have been shown in properly controlled studies to confer health benefits. Health effects of probiotics have been supported in numerous clinical trials as well as in high-quality systematic reviews and meta-analyses (Hoveyda et al. 2009; Hempel et al. 2012; Johnston et al. 2012; Hungin et al. 2013). Effects associated with probiotic consumption include improvement of digestion, stimulation of the immune system, alleviation of psychopathologies, decrease in antibiotic resistance, and anti-obesity effects, among others (Wieërs et al. 2020; Lagoumintzis and Patrinos 2023). However, as with all other discussed aspects of microbiota modulation, probiotic interventions often have different effects in different people (Maldonado-Gómez et al. 2016; Zmora et al. 2018).

The health benefits of probiotics can indeed be highly varied. They may be local to the gut or systemic, via host immune, endocrine, and nervous system mediators (Lebeer et al. 2018; Monteagudo-Mera et al. 2019; Plaza-Diaz et al. 2019). They can also be quite general, such as supporting a healthy gut microbiota and digestive tract, or more specific, such as reinforcing the immune system or reducing the risk of antibiotic-associated diarrhea. An administered probiotic may exert its effects directly—by providing a health-associated function that could be missing from the microbiome due to the lack of the administered microorganism or others capable of performing such function—or indirectly through its effects on the gut microbiome, either by enabling the cross-feeding of other beneficial microorganisms or by inhibiting the growth or metabolic activity of detrimental ones (Cunningham et al. 2021).

The health benefits of probiotics can also be exclusive to a single probiotic strain or shared among different species and genera. Most experts agree that supporting a healthy gut microbiota and digestive tract are general benefits that can be ascribed to the core group of well-studied probiotics, including various species of *Bifidobacterium* (e.g., *adolescentis, animalis, bifidum, breve,* and *longum*) and *Lactobacillus* (e.g., *acidophilus, casei, fermentum, gasseri, johnsonii, paracasei, plantarum, rhamnosus,* and *salivarius*). In fact, the general beneficial effect of these probiotics on the gut microbiota is likely to depend greatly on their capacity to promote a more generally favorable gut environment, by means of shared mechanisms such as SCFA production, competitive exclusion of pathogens, regulation of intestinal transit, and promotion of enterocyte turnover. In contrast, more specific health benefits—such as those deriving from bile acid metabolism or gut barrier reinforcement—tend to be restricted to certain probiotic species only. Further, health effects requiring the synthesis of specific bioactive molecules or the involvement of specific surface-associated effector molecules (e.g., pili, lipoteichoic acids, exopolysaccharides, surface-layer proteins) may be limited to just one or a few probiotic strains (Lebeer et al. 2018; Monteagudo-Mera et al. 2019; Cunningham et al. 2021). Strain-specific molecules are often involved in the generation of particular endocrine, immune, or neurologic effects (Hill et al. 2014).

Like prebiotics, probiotics can be consumed as part of functional foods or as food supplements. Importantly, probiotics do not have to be taken individually but can be ingested in preparations of defined microbial mixes. This approach has been termed "microbial ecosystems therapeutics" (Petrof et al. 2013) and is receiving substantial attention due to its ample potential for microbiota modulation. Indeed, such defined microbial consortia likely involve network interactions among their members, providing synergistic effects (Vázquez-Castellanos et al. 2019). For example, bile acid-metabolizing microbial consortia could be recruited to control elevated levels of primary bile acids (Cunningham et al. 2021). Probiotics can also be combined with prebiotic compounds, often aimed at promoting their growth or specific metabolic features; these probiotic–prebiotic mixtures have been termed "synbiotics" (Cunningham et al. 2021; Yadav et al. 2022). In addition, the live microbial cultures present in some fermented foods can be considered probiotics if they are proven to confer some health benefit. For instance, yogurt has been shown to improve lactose digestion in lactose maldigesters, and so the live bacteria in traditional yogurt cultures (*Streptococcus salivarius* subsp. *thermophilus* and *Lactobacillus delbrueckii* subsp. *bulgaricus*) can be considered probiotics (de Vrese et al. 2002; Guarner et al. 2005). Other fermented foods are natural sources of probiotic strains of lactic acid bacteria, and their intake has been associated with increased microbiota diversity and levels of butyrate-producing bacteria (Wieërs et al. 2020), as well as with important health benefits including reduced risk of type 2 diabetes and cardiovascular diseases (Cunningham et al. 2021). However, fermented foods with live microorganisms for which health benefits have not been proven should not be considered to contain probiotics.

Most common probiotics are lactic-acid-producing bacteria such as *Lactobacilli* and *Bifidobacteria* that were originally isolated from dairy products. With the development of affordable genome and metagenome sequencing, a better understanding of the composition and function of the human gut microbiota has suggested that certain individual members may represent excellent candidates for new probiotics. These are bacteria found to be associated with a robust gut microbiota composition and for which experimental work in animal models indicates that they are involved in important beneficial functions, such as mitigation of intestinal inflammation (Sokol et al. 2008), induction of immune regulation (Atarashi et al. 2013), or enhancement of the intestinal barrier function (Everard et al. 2013). These probiotic candidates include *Roseburia intestinalis* (Louis et al. 2007), *Faecalibacterium prausnitzii* (Sokol et al. 2008; Miquel et al. 2013), *Eubacterium hallii* (Vrieze et al. 2012), and *Akkermansia muciniphila* (Everard et al. 2013; Joyce and Gahan 2014), which offer physiological functions such as the production of butyrate, propionate, and other bioactive substances. The approval of these so-called "new-generation probiotics" for use in foods and food supplements is contingent on their being shown to meet all safety and efficacy requirements within the regulatory framework. In the case of *A. muciniphila*, it is already commercialized in a multi-species synbiotic, which also contains inulin, *Bifidobacterium longum* subsp. *infantis*, and other anaerobic bacteria (*Clostridium beijerinckii*, *Clostridium butyricum*, and *Anaerobutyricum hallii*). This synbiotic has been shown to improve glucose levels in individuals with type 2 diabetics (Perraudeau et al. 2020). In addition to "new-generation probiotics" isolated from the human microbiome, it is also possible to develop new probiotics that are genetically modified to express specific proteins or to produce specific metabolites that are beneficial to the host; however, the potential use of genetically modified organisms in humans remains a controversial issue that still needs to be resolved by regulatory authorities (Fan and Pedersen 2021).

Current developments in the field of microbiota modulation also encompass the use of bacterial products that could have effects similar to those of probiotics but that cannot be considered as such because they are not living microorganisms. These include dead microorganisms as well as microbial-derived fragments, metabolites, or other nonviable products. Fragments and metabolites from probiotic microorganisms that share some effects of their live counterparts have been termed "postbiotics" (Aguilar-Toalá et al. 2018). The health benefits of some of the new, promising gut microbiota isolates mentioned above may end up being exploited, at least partly, through the use of postbiotics. This is the case of *A. muciniphila,* for which studies in humans show that both live and pasteurized cells are safe to use and can improve several metabolic parameters (Depommier et al. 2019), while laboratory studies indicate that a purified membrane protein is sufficient to improve metabolism in obese and diabetic mice (Plovier et al. 2017).

1.7 Conclusions and Perspectives

Current advances in microbiome science have brought forward the potential to modulate the gut microbiota through a variety of dietary strategies. Also crucial in this respect will be advancing a detailed understanding of the chemical constituents of the foods we consume and their specific interactions within the gut microbiome.

When combined with the use of probiotics chosen from an ever-expanding range of microorganisms whose specific functions and health-promoting capabilities are increasingly well understood, dietary strategies clearly hold great promise for counteracting the deleterious effects of the Westernized lifestyle on the microbiome and health.

However, the field of gut microbiome modulation through dietary approaches also faces important challenges. Some of these are conceptual, such as the inherent difficulties in defining a healthy microbiome and measuring improvement toward a healthy status—especially in light of the tremendous interindividual variability in microbiome composition. This variability, together with the many other genetic and lifestyle factors that differ across individuals, tends to render the effects of dietary interventions highly personalized. This means that food-based interventions to modulate the gut microbiome will likely have to be tailored to specific individuals, taking into account their microbiome, health status, lifestyle habits, and other factors. Dietary strategies aimed at gut microbiome modulation—including the use of specific prebiotics, probiotics, and other food supplements—are already being targeted to specific population segments, such as the elderly or people suffering from specific health problems, including obesity (Dahiya et al. 2017). Further stratification and personalization will undoubtedly follow, transforming personalized nutrition into an important component of the current drive towards personalized medicine and well-being.

Other challenges in the field are technical. For example, an enormous amount of fundamental and translational research is needed in order to optimize issues of dosage and delivery of prebiotics, probiotics, and other food supplements. Also, the challenge of studying and compiling the composition of the multitude of foods consumed by humans is daunting and will necessitate the joint action of numerous actors around the world, as well as the constant improvement of food composition databases. Clearly, the integration of all the complex strands of data coming from microbiome and food science into actual personalized dietary recommendations will require sophisticated and powerful computational approaches. Chapters 6 and 7 will address some of the current advances in metabolic modeling of the gut microbiome for personalized nutrition and the use of computational applications to incorporate microbiome information into individually generated dietary recommendations.

Finally, the diversification in the types, mechanisms of action, and potential health-related applications of novel functional foods, prebiotics, probiotics, and other food supplements will require the concerted evolution of regulatory policies to govern their use and health claims. The development of such regulatory frameworks will involve numerous national and international agencies and will likely result in nonidentical regulations across different countries. With these frameworks in place, the science of gut microbiome modulation through foods and food supplements should have the capacity to provide great personal and societal benefits in the hands of responsible consumers, nutritionists, and medical doctors.

References

Adak A, Khan MR (2019) An insight into gut microbiota and its functionalities. Cell Mol Life Sci 76:473–493

Aguilar-Toalá J et al (2018) Postbiotics: an evolving term within the functional foods field. Trends Food Sci Technol 75:105–114

Anhê FF et al (2015) A polyphenol-rich cranberry extract protects from diet-induced obesity, insulin resistance and intestinal inflammation in association with increased *Akkermansia* spp. population in the gut microbiota of mice. Gut 64:872–883

Atarashi K et al (2013) Treg induction by a rationally selected mixture of clostridia strains from the human microbiota. Nature 500:232–236

Bailey LC et al (2014) Association of antibiotics in infancy with early childhood obesity. JAMA Pediatr 168:1063–1069. https://doi.org/10.1001/jamapediatrics.2014.1539

Barone M et al (2022) Gut microbiome–micronutrient interaction: the key to controlling the bioavailability of minerals and vitamins? Biofactors 48:307. https://doi.org/10.1002/BIOF.1835

Barratt MJ et al (2017) The gut microbiota, food science, and human nutrition: a timely marriage. Cell Host Microbe 22:134–141

Bielik V, Kolisek M (2021) Bioaccessibility and bioavailability of minerals in relation to a healthy gut microbiome. Int J Mol Sci 22. https://doi.org/10.3390/IJMS22136803

Braune A, Engst W, Blaut M (2015) Identification and functional expression of genes encoding flavonoid O-and C-glycosidases in intestinal bacteria. Environ Microbiol 18:2117–2129

Brown AJ et al (2003) The orphan G protein-coupled receptors GPR41 and GPR43 are activated by propionate and other short chain carboxylic acids. J Biol Chem 278:11312–11319

Cardona F et al (2013) Benefits of polyphenols on gut microbiota and implications in human health. J Nutr Biochem 24:1415–1422

Chatelier EL et al (2013) Richness of human gut microbiome correlates with metabolic markers. Nature 500:541–546

Chen Y et al (2020) Gut microbiota and short chain fatty acid composition as affected by legume type and processing methods as assessed by simulated *in vitro* digestion assays. Food Chem 32: 126040

Claesson MJ et al (2012) Gut microbiota composition correlates with diet and health in the elderly. Nature 488:178–184. https://doi.org/10.1038/NATURE11319

Clemente JC et al (2012) The impact of the gut microbiota on human health: an integrative view. Cell 148(6):1258–1270

Clifford MN (2004) Diet-derived phenols in plasma and tissues and their implications for health. Planta Med 70:1103–1114

Cunningham M et al (2021) Shaping the future of probiotics and prebiotics. Trends Microbiol 29: 667–685. https://doi.org/10.1016/j.tim.2021.01.003

Dahiya DK et al (2017) Gut microbiota modulation and its relationship with obesity using prebiotic fibers and probiotics: a review. Front Microbiol 8:563. https://doi.org/10.3389/fmicb.2017.00563

David LA et al (2014) Diet rapidly and reproducibly alters the human gut microbiome. Nature 505: 559–563. https://doi.org/10.1038/nature12820

De Filippis F et al (2016) High-level adherence to a Mediterranean diet beneficially impacts the gut microbiota and associated metabolome. Gut 65:1812–1821. https://doi.org/10.1136/gutjnl-2015-309957

De Filippo C et al (2010) Impact of diet in shaping gut microbiota revealed by a comparative study in children from Europe and rural Africa. Proc Natl Acad Sci USA 107:14691–14696

De Vadder F et al (2014) Microbiota-generated metabolites promote metabolic benefits via gut-brain neural circuits. Cell 156:84–96

De Vadder F et al (2016) Microbiota-produced succinate improves glucose homeostasis via intestinal gluconeogenesis. Cell Metab 24:151–157

de Vos WM et al (2022) Gut microbiome and health: mechanistic insights. Gut 71:1020–1032 de Vrese M et al (2002) Probiotics – compensation for lactase insufficiency. Am J Clin Nutr 73: 421S–429S

Delcour JA et al (2016) Prebiotics, fermentable dietary fiber, and health claims. Adv Nutr 7:1–4

Delmée E et al (2006) Relation between colonic proglucagon expression and metabolic response to oligofructose in high fat diet-fed mice. Life Sci 79:1007–1013. https://doi.org/10.1016/j.lfs.2006.05.013

Depommier C et al (2019) Supplementation with *Akkermansia muciniphila* in overweight and obese human volunteers: a proof-of-concept exploratory study. Nat Med 25:1096–1103

Dewulf EM et al (2013) Insight into the prebiotic concept: lessons from an exploratory, double blind intervention study with inulin-type fructans in obese women. Gut 62:1112–1121

Duda-Chodak A et al (2015) Interaction of dietary compounds, especially polyphenols, with the intestinal microbiota: a review. Eur J Nutr 54:325–341

Duenas M et al (2015) A survey of modulation of gut microbiota by dietary polyphenols. Biomed Res Int 2015:1

Duncan SH et al (2007) Reduced dietary intake of carbohydrates by obese subjects results in decreased concentrations of butyrate and butyrate-producing bacteria in feces. Appl Environ Microbiol 73:1073–1078

Everard A et al (2013) Cross-talk between Akkermansia muciniphila and intestinal epithelium controls diet-induced obesity. Proc Natl Acad Sci USA 110:9066–9071

Fan Y, Pedersen O (2021) Gut microbiota in human metabolic health and disease. Nat Rev Microbiol 19(1):55–71. https://doi.org/10.1038/s41579-020-0433-9

Flamm G et al (2001) Inulin and oligofructose as dietary fiber: a review of the evidence. Crit Rev Food Sci Nutr 41(5):353–362

Flint HJ et al (2012) Microbial degradation of complex carbohydrates in the gut. Gut Microbes 3: 289–306. https://doi.org/10.4161/gmic.19897

Francino MP (2016) Antibiotics and the human gut microbiome: Dysbioses and accumulation of resistances. Front Microbiol 6:1543. https://doi.org/10.3389/fmicb.2015.01543

Francino MP (2017) The gut microbiome and metabolic health. Curr Nutr Rep 6:16–23. https://doi.org/10.1007/s13668-017-0190-1

Francois IEJA et al (2012) Effects of a wheat bran extract containing arabinoxylan oligosaccharides on gastrointestinal health parameters in healthy adult human volunteers: a double-blind, randomised, placebo-controlled, cross-over trial. Br J Nutr 108:2229–2242

Frost G et al (2014) The short-chain fatty acid acetate reduces appetite via a central homeostatic mechanism. Nat Commun 5:3611

Gibson GR, Roberfroid MB (1995) Dietary modulation of the human colonic microbiota: introducing the concept of prebiotics. J Nutr 125:1401–1412

Gibson GR et al (2017) Expert consensus document: the international scientific Association for Probiotics and Prebiotics (ISAPP) consensus statement on the definition and scope of prebiotics. Nat Rev Gastroenterol Hepatol 14:491–502

Gill CIR et al (2010) Profiling of phenols in human fecal water after raspberry supplementation. J Agric Food Chem 58:10389–10395

Guarner F et al (2005) Should yoghurt cultures be considered probiotic? Br J Nutr 93:783–786

Gutiérrez-Díaz I et al (2016) Mediterranean diet and faecal microbiota: a transversal study. Food Funct 7(5):2347–2356. https://doi.org/10.1039/c6fo00105j

Hansen MEB et al (2019) Population structure of human gut bacteria in a diverse cohort from rural Tanzania and Botswana. Genome Biol 20:16

Haro C et al (2017) Consumption of two healthy dietary patterns restored microbiota dysbiosis in obese patients with metabolic dysfunction. Mol Nutr Food Res 61:1700300. https://doi.org/10.1002/mnfr.201700300

Healey GR et al (2017) Interindividual variability in gut microbiota and host response to dietary interventions. Nutr Rev 75:1059–1080. https://doi.org/10.1093/nutrit/nux062

Heiman ML, Greenway FL (2016) A healthy gastrointestinal microbiome is dependent on dietary diversity. Mol Metab 5:317–320

Hempel S et al (2012) Probiotics for the prevention and treatment of antibiotic-associated diarrhea: a systematic review and meta-analysis. JAMA 307:1959–1969

Herman DR et al (2019) Dietary habits of two to nine-year-old American children are associated with gut microbiome composition. J Acad Nutr Diet 120:517. https://doi.org/10.1016/J.JAND.2019.07.024

Hill C et al (2014) Expert consensus document. The international scientific Association for Probiotics and Prebiotics consensus statement on the scope and appropriate use of the term probiotic. Nat Rev Gastroenterol Hepatol 11:506–514

Hill P, Muir JG, Gibson PR (2017) Controversies and recent developments of the low-FODMAP diet. Gastroenterol Hepatol 13:36–45

Hoveyda N et al (2009) A systematic review and meta-analysis: probiotics in the treatment of irritable bowel syndrome. BMC Gastroenterol 9:15

Human Microbiome Project Consortium (2012) Structure, function and diversity of the healthy human microbiome. Nature 486:207–214

Hungin AP et al (2013) Systematic review: probiotics in the management of lower gastrointestinal symptoms in clinical practice – an evidence-based international guide. Aliment Pharmacol Ther 38:864–886

Hviid A, Svanstrom H, Frisch M (2011) Antibiotic use and inflammatory bowel diseases in childhood. Gut 60(1):49–54. https://doi.org/10.1136/gut.2010.219683

Jernberg C et al (2007) Long-term ecological impacts of antibiotic administration on the human intestinal microbiota. ISME J 1:56–66. https://doi.org/10.1038/ismej.2007.3

Johnson AJ et al (2019) Daily sampling reveals personalized diet-microbiome associations in humans. Cell Host Microbe 25:789–802.e5. https://doi.org/10.1016/J.CHOM.2019.05.005

Johnston BC et al (2012) Probiotics for the prevention of *Clostridium difficile*-associated diarrhea: a systematic review and meta-analysis. Ann Intern Med 157:878–888

Jones BV et al (2008) Functional and comparative metagenomic analysis of bile salt hydrolase activity in the human gut microbiome. Proc Natl Acad Sci USA 105:13580e13585

Joyce SA, Gahan CG (2014) The gut microbiota and the metabolic health of the host. Curr Opin Gastroenterol 30:120–127

Kimble R et al (2023) Effects of a mediterranean diet on the gut microbiota and microbial metabolites: a systematic review of randomized controlled trials and observational studies. Crit Rev Food Sci Nutr 63(27):8698–8719. https://doi.org/10.1080/10408398.2022.2057416

Kovatcheva-Datchary P et al (2015) Dietary fiber-induced improvement in glucose metabolism is associated with increased abundance of *Prevotella*. Cell Metab 22:971–982

Kumar Singh A et al (2019) Beneficial effects of dietary polyphenols on gut microbiota and strategies to improve delivery efficiency. Nutrients 11:2216

Lagoumintzis G, Patrinos GP (2023) Triangulating nutrigenomics, metabolomics and microbiomics toward personalized nutrition and healthy living. Hum Genomics 17:109. https://doi.org/10.1186/s40246-023-00561-w

Lampe JW (2009) Is equol the key to the efficacy of soy foods? Am J Clin Nutr 89:1664S–1667S

Landete JM et al (2015) Bioactivation of phytoestrogens: intestinal bacteria and health. Crit Rev Food Sci Nutr 56:1826–1843

Larsen N et al (2010) Gut microbiota in human adults with type 2 diabetes differs from non-diabetic adults. PLoS One 5:e9085

Lebeer S et al (2018) Identification of probiotic effector molecules: present state and future perspectives. Curr Opin Biotechnol 49:217–223

Leeming ER et al (2019) Effect of diet on the gut microbiota: rethinking intervention duration. Nutrients 11. https://doi.org/10.3390/NU11122862

Lerma-Aguilera AM (2024) Evaluation of the *in vitro* response of the intestinal microbiota to various foods and of the impact of a personalized nutritional intervention. Doctoral Thesis, University of Valencia

Lerma-Aguilera AM et al (2024) Effects of different foods and cooking methods on the gut microbiota: an in vitro approach. Front Microbiol 14:1334623. https://doi.org/10.3389/fmicb.2023.1334623

Louis P, Flint HJ (2017) Formation of propionate and butyrate by the human colonic microbiota. Environ Microbiol 19(1):29–41

Louis P et al (2007) Understanding the effects of diet on bacterial metabolism in the large intestine. J Appl Microbiol 102:1197–1208

Lv Z et al (2016) Vitamin A deficiency impacts the structural segregation of gut microbiota in children with persistent diarrhea. J Clin Biochem Nutr 59:113–121. https://doi.org/10.3164/jcbn.15-148

Maier TV et al (2017) Impact of dietary resistant starch on the human gut microbiome, metaproteome, and metabolome. MBio 8:e01343–e01317. https://doi.org/10.1128/mBio.01343-17

Maldonado-Gómez MX et al (2016) Stable engraftment of *Bifidobacterium longum* AH1206 in the human gut depends on individualized features of the resident microbiome. Cell Host Microbe 20:515–526

Manach C et al (2004) Polyphenols: food sources and bioavailability. Am J Clin Nutr 79:727–747

Marín L et al (2015) Bioavailability of dietary polyphenols and gut microbiota metabolism: antimicrobial properties. Biomed Res Int 2015:905215

McDonald D et al (2018) American gut: an open platform for citizen science microbiome research. mSystems 3:10-1128

Meslier V et al (2020) Mediterranean diet intervention in overweight and obese subjects lowers plasma cholesterol and causes changes in the gut microbiome and metabolome independently of energy intake. Gut 69:1258–1268

Miquel S et al (2013) *Faecalibacterium prausnitzii* and human intestinal health. Curr Opin Microbiol 16:255–261

Mitsou EK et al (2017) Adherence to the Mediterranean diet is associated with the gut microbiota pattern and gastrointestinal characteristics in an adult population. Br J Nutr 117(12):1645–1655. https://doi.org/10.1017/S0007114517001593

Molino S et al (2021) Enrichment of food with tannin extracts promotes healthy changes in the human gut microbiota. Front Microbiol 12:625782. https://doi.org/10.3389/fmicb.2021.625782

Molino S et al (2022) Evaluation of the effects of a short supplementation with tannins on the gut microbiota of healthy subjects. Front Microbiol 13:848611. https://doi.org/10.3389/FMICB.2022.848611

Molino S et al (2023) Evaluation of tannin-delivery approaches for gut microbiota modulation: comparison of pectin-based microcapsules and unencapsulated extracts. J Agric Food Chem 71:13988–13999. https://doi.org/10.1021/ACS.JAFC.3C02949

Monteagudo-Mera A et al (2019) Adhesion mechanisms mediated by probiotics and prebiotics and their potential impact on human health. Appl Microbiol Biotechnol 103:6463–6472

Moriki D et al (2022) The role of the gut microbiome in cow's Milk allergy: a clinical approach. Nutrients 14:4537. https://doi.org/10.3390/nu14214537

Navajas-Porras B et al (2020) Effect of cooking methods on the antioxidant capacity of plant foods submitted to in vitro digestion–fermentation. Antioxidants 9:1–23. https://doi.org/10.3390/antiox9121312

Obregon-Tito AJ et al (2015) Subsistence strategies in traditional societies distinguish gut microbiomes. Nat Commun 6:6505

Ott SJ et al (2004) Reduction in diversity of the colonic mucosa associated bacterial microflora in patients with active inflammatory bowel disease. Gut 53:685–693

Panagiotakos DB, Pitsavos C, Stefanadis C (2006) Dietary patterns: a Mediterranean diet score and its relation to clinical and biological markers of cardiovascular disease risk. Nutr Metab Cardiovasc Dis 16:559–568

Pasolli E et al (2019) Extensive unexplored human microbiome diversity revealed by over 150,000 genomes from metagenomes spanning age, geography, and lifestyle. Cell 176:649–662

Pedersen HK et al (2016) Human gut microbes impact host serum metabolome and insulin sensitivity. Nature 535:376–381

Pérez-Burillo S et al (2018) Effect of food thermal processing on the composition of the gut microbiota. J Agric Food Chem 66:11500–11509. https://doi.org/10.1021/acs.jafc.8b04077

Perraudeau F et al (2020) Improvements to postprandial glucose control in subjects with type 2 diabetes: a multicenter, double blind, randomized placebo-controlled trial of a novel probiotic formulation. BMJ Open Diabetes Res Care 8:e001319

Petrof EO et al (2013) Microbial ecosystems therapeutics: a new paradigm in medicine? Benef Microbes 4:53–65

Plaza-Diaz J et al (2019) Mechanisms of action of probiotics. Adv Nutr 10:S49–S66

Plovier H et al (2017) A purified membrane protein from *Akkermansia muciniphila* or the pasteurized bacterium improves metabolism in obese and diabetic mice. Nat Med 23:107–113

Quartieri A et al (2016) Detection of novel metabolites of flaxseed lignans *in vitro* and *in vivo*. Mol Nutr Food Res 60:1590–1601

Reddel S, Putignani L, Del Chierico F (2019) The impact of low-FODMAPs, gluten-free, and ketogenic diets on gut microbiota modulation in pathological conditions. Nutrients 11:373. https://doi.org/10.3390/nu11020373

Reichardt N et al (2014) Phylogenetic distribution of three pathways for propionate production within the human gut microbiota. ISME J 8:1323–1335

Rey-Mariño A, Francino MP (2022) Nutrition, gut microbiota, and allergy development in infants. Nutrients 14:4316. https://doi.org/10.3390/nu14204316

Rothschild D et al (2018) Environment dominates over host genetics in shaping human gut microbiota. Nature 555:210–215. https://doi.org/10.1038/NATURE25973

Rowland I et al (2003) Bioavailability of phyto-oestrogens. Br J Nutr 89(Suppl 1):S45–S58

Rowland I et al (2018) Gut microbiota functions: metabolism of nutrients and other food components. Eur J Nutr 57:1–24. https://doi.org/10.1007/s00394-017-1445-8

Russell WR et al (2008) Anti-inflammatory implications of the microbial transformation of dietary phenolic compounds. Nutr Cancer 60:636–642

Rutz A et al (2022) The LOTUS initiative for open knowledge management in natural products research. elife 11:e70780. https://doi.org/10.7554/eLife.70780

Saari A et al (2015) Antibiotic exposure in infancy and risk of being overweight in the first 24 months of life. Pediatrics 135:617–626. https://doi.org/10.1542/peds.2014-3407

Sanz Y (2010) Effects of a gluten-free diet on gut microbiota and immune function in healthy adult humans. Gut Microbes 1:135–137. https://doi.org/10.4161/gmic.1.3.11868

Sayin SI et al (2013) Gut microbiota regulates bile acid metabolism by reducing the levels of tauro-beta-muricholic acid, a naturally occurring FXR antagonist. Cell Metab 17:225–235

Schnorr SL et al (2014) Gut microbiome of the Hadza hunter-gatherers. Nat Commun 5:3654

Scott KP et al (2013) The influence of diet on the gut microbiota. Pharmacol Res 69(1):52–60

Sekirov I et al (2010) Gut microbiota in health and disease. Physiol Rev 90:859–904

Shankar V et al (2017) Differences in gut metabolites and microbial composition and functions between Egyptian and U.S. children are consistent with their diets. mSystems 2:e00169–e00116

Smits SA et al (2017) Seasonal cycling in the gut microbiome of the Hadza hunter-gatherers of Tanzania. Science 357:802–806. https://doi.org/10.1126/science.aan4834

Sofi F et al (2008) Adherence to Mediterranean diet and health status: meta-analysis. BMJ 337: a1344

Sokol H et al (2008) *Faecalibacterium prausnitzii* is an anti-inflammatory commensal bacterium identified by gut microbiota analysis of Crohn disease patients. Proc Natl Acad Sci USA 105: 16731–16736

Sonnenburg JL, Bäckhed F (2016) Diet–microbiota interactions as moderators of human metabolism. Nature 535:56–64. https://doi.org/10.1038/nature18846

Sonnenburg ED et al (2016) Diet-induced extinctions in the gut microbiota compound over generations. Nature 529:212–215

Steliou K et al (2012) Butyrate histone deacetylase inhibitors. Biores Open Access 1:192–198

Tazoe H et al (2008) Roles of short-chain fatty acids receptors, GPR41 and GPR43 on colonic functions. J Physiol Pharmacol 59(Suppl 2):251–262
Teng Y et al (2018) Plant-derived Exosomal MicroRNAs shape the gut microbiota. Cell Host Microbe 24:637–652. https://doi.org/10.1016/j.chom.2018.10.001
Tett A et al (2019) The *Prevotella copri* complex comprises four distinct clades underrepresented in westernized populations. Cell Host Microbe 26:666–679. https://doi.org/10.1016/j.chom.2019.08.018
Tomas-Barberan F et al (2014) *In vitro* transformation of chlorogenic acid by human gut microbiota. Mol Nutr Food Res 58:1122–1131
Trasande L et al (2012) Infant antibiotic exposures and early-life body mass. Int J Obes 37:16–23. https://doi.org/10.1038/ijo.2012.132
Tuohy KM et al (2001) A human volunteer study on the prebiotic effects of HP-inulin – Faecal bacteria enumerated using fluorescent in situ hybridisation (FISH). Anaerobe 7:113–118. https://doi.org/10.1006/ANAE.2001.0368
Turnbaugh PJ et al (2009) A core gut microbiome in obese and lean twins. Nature 457:480–484
Vandeputte D et al (2017) Prebiotic inulin-type fructans induce specific changes in the human gut microbiota. Gut 66:1968. https://doi.org/10.1136/gutjnl-2016-313271
Vangay P et al (2018) US immigration westernizes the human gut microbiome. Cell 175:962–972
Vázquez-Castellanos JF et al (2019) Design of synthetic microbial consortia for gut microbiota modulation. Curr Opin Pharmacol 49:52–59
Vrieze A et al (2012) Transfer of intestinal microbiota from lean donors increases insulin sensitivity in individuals with metabolic syndrome. Gastroenterology 143:913–916
Walker AW et al (2011) Dominant and diet-responsive groups of bacteria within the human colonic microbiota. ISME J 5:220–230. https://doi.org/10.1038/ismej.2010.118
Wang Z et al (2011) Gut flora metabolism of phosphatidylcholine promotes cardiovascular disease. Nature 472:57–63
Wang Z et al (2015) Non-lethal inhibition of gut microbial trimethylamine production for the treatment of atherosclerosis. Cell 163:1585–1595
Wang S et al (2020) Rational use of prebiotics for gut microbiota alterations: specific bacterial phylotypes and related mechanisms. J Funct Foods 66:103838
Wieërs G et al (2020) How probiotics affect the microbiota. Front Cell Infect Microbiol 9:454. https://doi.org/10.3389/fcimb.2019.00454
Willett WC et al (1995) Mediterranean diet pyramid: a cultural model for healthy eating. Am J Clin Nutr 61(Suppl. 6):S1402–S1406
Wu GD et al (2011) Linking long-term dietary patterns with gut microbial enterotypes. Science 334: 105–108. https://doi.org/10.1126/science.1208344
Yadav MK et al (2022) Probiotics, prebiotics and synbiotics: safe options for next-generation therapeutics. Appl Microbiol Biotechnol 106:505–521. https://doi.org/10.1007/s00253-021-11646-8
Yatsunenko T et al (2012) Human gut microbiome viewed across age and geography. Nature 486: 222–227
Zeevi D et al (2015) Personalized nutrition by prediction of glycemic responses. Cell 163:1079–1094. https://doi.org/10.1016/j.cell.2015.11.001
Zinöcker M, Lindseth I (2018) The Western diet–microbiome-host interaction and its role in metabolic disease. Nutrients 10:365. https://doi.org/10.3390/nu10030365
Zmora N et al (2018) Personalized gut mucosal colonization resistance to empiric probiotics is associated with unique host and microbiome features. Cell 174:1388–1405

Effect of Food Processing on Gut Microbiota Composition and Functionality

2

Adriana Delgado-Osorio, Silvia Pastoriza, and José Ángel Rufián Henares

Abstract

Food processing transforms the physical and chemical properties of food, potentially modifying their effect on gut microbiota composition and functionality. This chapter explores how thermal food processing methods, both at the household and industrial levels, impact the gut microbiota. Research shows that the impact of thermal processing strongly depends on the food type, thermal processing method, and specific measurements of composition and functionality. For gut microbiota composition and short-chain fatty acid (SCFA) production, thermal processing tends to have a more positive effect on carbohydrate-rich foods, while lipid-rich foods yield the least favorable outcomes. Thermal processing also shows positive effects on the bioaccessibility and microbial metabolism of (poly)phenols. Thermal processing can also enhance the gut microbiota's ability to improve the bioaccessibility of minerals and carotenoids, though it is generally detrimental to the release of antioxidant capacity. Microwaving has proven to be one of the most effective thermal processing methods for promoting gut microbiota composition and functionality. The effect of compounds generated during thermal processing (e.g., Maillard reaction products) on the gut microbiota remains inconclusive. Thermal food processing could also be a key factor to consider in personalized nutrition for enhancing the composition and functionality of the gut microbiota at the individual level.

A. Delgado-Osorio · S. Pastoriza · J. Á. Rufián Henares (✉)
Department of Nutrition & Food Science, Faculty of Pharmacy, University of Granada, Granada, Andalucía, Spain
e-mail: jarufian@ugr.es

M. P. Francino, J. Á. Rufián Henares (eds.), *The Microbiome and Personalized Nutrition*, The Microbiomes of Humans, Animals, Plants, and the Environment 6,
https://doi.org/10.1007/978-3-031-93902-0_2

Keywords

Food thermal processing · Gut microbiota · Metabolites · Short-chain fatty acids · (Poly)phenols · Antioxidant capacity · Maillard reaction

2.1 Introduction

Food processing plays a significant role in the way we consume food. Among all the methods of food processing, thermal food processing is the most widespread. Thermal food processing involves raising the temperature of a food through direct or indirect heat and is prevalent both domestically and industrially. At the household level, methods such as boiling, frying, grilling, baking, roasting, steaming, toasting, and microwaving are common. Industrial processes include canning, pasteurization, sterilization, and ultra-high temperature treatments. Thermal food processing offers numerous benefits, including improved food safety, palatability, and convenience. However, it also presents some drawbacks, such as the potential loss of beneficial compounds and the formation of undesirable ones (van Boekel et al. 2010).

Thermal treatment causes significant physical and chemical changes in food. Heat alters macronutrients like starch and proteins, impacting their digestibility and therefore their bioavailability. For example, heat generally improves starch digestibility but can also increase the formation of resistant starch with reduced digestibility (van Boekel et al. 2010). Proteins may become more digestible through denaturation, but heat can also lead to glycation and oxidation reactions that favor cross-linked structures that are harder to digest (Ashkar and Wu 2023; Wu et al. 2022). Heat can also induce the hydrolysis of dietary fiber, therefore disrupting the plant cell walls and leading to the release of elements and compounds that were bound to the cell walls or confined within the cells.

Phytochemicals can also be altered by thermal processing. (Poly)phenols can be degraded by high temperature and undergo isomerization and transesterification reactions, which has an impact on their bioaccessibility and antioxidant capacity (Neilson and Ferruzzi 2011). Other alterations related to thermal processing are the formation of new compounds. Maillard reaction products (e.g., furfurals, melanoidins, and advanced glycation end products), acrylamide, and heterocyclic amines are examples of compounds produced after thermal processing, with both positive (e.g., antioxidant) and negative (e.g., mutagenic) health effects (Tuohy et al. 2006). All these changes affect the bioavailability of food elements and compounds, altering what reaches the colon, where gut microbiota is most abundant (Rinninella et al. 2019).

The human gut microbiota is a complex community of microorganisms composed mainly of bacteria from the phyla Firmicutes and Bacteroidetes. Two bacterial families belonging to the Firmicutes, Lachnospiraceae and Ruminococcaceae, each comprise 20%–30% of colonic microbiotas (Tannock 2023). The phyla Actinobacteria, Proteobacteria, and Verrucomicrobia are also represented. At lower taxonomical levels, the gut microbiota's complexity varies greatly among individuals. However, the gut microbiota shows high functional redundancy, meaning that some functions can be conferred by multiple bacteria. Therefore, when assessing the impact of a specific factor (e.g., food thermal processing) on the gut microbiota, it is essential to identify specific changes in both its composition and functionality.

The functionality of the gut microbiota is diverse, with one of its most prevalent actions being the metabolism of compounds unabsorbed in the upper gastrointestinal tract. The human gut microbiota can metabolize these compounds through various pathways, resulting in the production of a wide range of metabolites.

The gut microbiota metabolizes nondigestible carbohydrates and proteins into beneficial short-chain fatty acids (SCFA), mainly acetic, propionic, and butyric acids. The gut microbiota can also transform phytochemicals, like (poly)phenols, into beneficial metabolites, by hydrolysis, cleavage, and reduction reactions (Tomás-Barberán and Espín 2019). Also, the gut microbiota metabolizes fiber, releasing fiber-bound or entrapped elements and compounds, like minerals and phytochemicals, thus incrementing their bioaccessibility. The gut microbiota can also modify the antioxidant capacity of food, by producing, degrading, or releasing antioxidant elements and compounds from the food matrix.

Studying the gut microbiota involves examining its composition and functionality. Human studies are the preferred option, but due to their challenges, researchers often use in vivo animal models or in vitro fermentation models that mimic intestinal conditions. Regarding the compositional assessment of the gut microbiota, 16S rRNA gene sequencing is the most widespread technique, while functionality can be assessed through gene expression and microbial metabolite production (e.g., SCFA and (poly)phenol metabolites). Nevertheless, gut microbiota functionality can also be indirectly assessed by evaluating the effect on a specific measurement (e.g., bioaccessibility of elements and compounds and antioxidant capacity) after incubating food with gut microbiota.

To our knowledge, no review has yet been conducted on the effects of thermal food processing on gut microbiota composition and functionality. Therefore, this book chapter aims to evaluate the impact of thermal food processing on gut microbiota composition and functionality and explore the impact of compounds generated by food thermal processing on the gut microbiota.

2.2 Effect of Food Processing on Gut Microbiota Composition

Few studies have compared diets that are thermally processed with those that are not. These approaches are particularly valuable because they evaluate the entire diet rather than focusing on a single food item providing a more realistic representation of food consumption. To our knowledge, the only attempt to understand the effects of thermal food processing on humans is the study conducted by Carmody et al. (2019). In two three-day interventions in counterbalanced order, researchers compared the effects of matched raw and cooked plant-based diets on gut microbiota composition (Carmody et al. 2019). Although treatment responses differed by participant, cooking significantly impacted gut microbial β-diversity, with detectable changes within 48 h on cooked diets, persisting until day five. The raw diet showed a delayed response, becoming significant at 72 h and lasting for only 24 h. Most diet-responsive operational taxonomic units (OTUs) were uniquely enriched on either raw or cooked diets, spanning multiple taxonomic groups.

Other studies have investigated the effects of thermally processed diets on animals, producing mixed results. However, a trend of reduced α-diversity due to thermal processing was observed. In one of the studies, researchers fed beagle dogs raw, pasteurized, or high-temperature sterilized diets and analyzed their effects on gut microbiota composition (Cai et al. 2022). High-temperature sterilization decreased both the number of observed species and the Chao1 species richness index in the dogs. At least 14 genera were significantly affected by thermal processing. In particular, the abundance of *Alloprevotella*, *Escherichia–Shigella*, and *Turicibacter* was lower in beagles fed the raw diet, whereas *Streptococcus*, *Collinsella*, *Bacteroides*, and *Ruminococcus gnavus* were more abundant following the high-temperature sterilization diet. On the other hand, beagles fed the pasteurized diet exhibited the highest abundance of *Lactococcus*.

The other study performed in animals analyzed the effects of a low-fat diet, a high-fat diet, and a heat-treated high-fat diet on the gut microbiota composition of the cecum of ApoE$^{-/-}$ mice (Marungruang et al. 2016). Heat-treated high-fat diet lowered the α-diversity of the gut microbiota significantly. In regard to community composition, nonparametric analysis revealed significant differences between the microbial communities of the three groups. At the phylum level, the heat-treated high-fat diet group had a significantly higher abundance of Firmicutes and lower abundance of Bacteroidetes compared to the low-fat and high-fat diet groups. Other significant results observed in the heat-treated high-fat diet group were the lower Rikenellaceae and higher *Allobaculum* abundances.

Research has also examined multiple food items simultaneously under the same experimental conditions providing insights into how different thermal food processing methods can alter each specific item. Also, by studying a variety of foods together, these studies offer a broader and more comprehensive understanding than those focusing on a single food item. An evaluation of 159 combinations of foods and thermal processing methods after in vitro digestion and fermentation with the feces of three healthy individuals indicated that the effects of thermal processing were highly individualized and dependent on the specific food, although some general trends were observed across different food categories and individuals (Lerma-Aguilera et al. 2024). Notably, frying consistently resulted in a lower abundance of Ruminococcaceae UBA1819, a butyric acid producer, compared to roasted foods across all categories and individuals. Frying also led to significant differences in taxon abundance compared to other methods, particularly within fruits, vegetables, and bread, generally showing higher abundances of genera within the Lachnospiraceae family and lower abundances within the Ruminococcaceae family. Similar trends were observed for grilling. Boiling also produced significant differences compared to other methods in various food categories, particularly fish and meat, generally leading to a lower abundance of many Firmicutes genera.

Additional studies have considered fewer food items but have incorporated an important factor when analyzing the effect of thermal food processing on the gut microbiota: the degree of thermal processing. One of these investigations involved in vitro digestion and fermentation of eggplant, garlic, and onion, all fried, grilled, or roasted over two processing times (normal and well done) (Rajakaruna et al. 2024). The type of vegetable had the most significant impact on gut microbiota structure, followed by processing method and duration. Furosine, a Maillard reaction product and an indicator of thermal processing degree, was only weakly associated with gut microbiota structure. The prebiotic index revealed that roasting produced the most beneficial microbiota structure, due to the support of *Akkermansia* and *Bifidobacterium* and inhibition of Enterobacteriaceae. Well-done samples maintained higher levels of *Akkermansia* and *Bifidobacterium*, resulting in a higher prebiotic index ratio compared to normally cooked samples.

Another work considering the degree of thermal processing submitted different processed foods to in vitro digestion and fermentation: wheat bread (raw and lightly and highly toasted), chicken (boiled, fried, grilled, and roasted), red pepper (raw, fried, and roasted), chickpeas (boiled and grilled), and banana (raw, roasted, and fried) (Pérez-Burillo et al. 2018). Raw banana and bread as well as boiled chickpeas showed high abundances in beneficial bacteria such as *Roseburia*, *Ruminococcus*, *Bifidobacterium*, *Dialister*, *Collinsella*, or *Barnesiella*. Toasted bread resulted in lower levels of beneficial bacteria, which were even lower in the highly toasted form. Few differences between cooking methods were found for chicken and pepper. Significant positive correlations were observed between certain bacteria and Maillard reaction products (furosine, hydroxymethylfurfural, and furfural), which could indicate that the Maillard reaction fosters the growth of these bacteria. However, one negative correlation was detected, indicating that these compounds could act as inhibitors too.

In contrast to the aforementioned studies, which assess diets or multiple food items simultaneously, some research has taken a smaller-scale approach by investigating how thermal processing of individual food items, tested either alone or included in diets, affects gut microbiota composition. Various types of food items have been examined, and we have categorized the results based on the similarity of their macronutrient content.

2.2.1 Carbohydrate-Rich Foods

Regarding the impact of thermally processing carbohydrate-rich foods on gut microbiota composition, mixed results were obtained. Nevertheless, thermal treatment, particularly methods like microwaving and steaming, was generally more beneficial than detrimental.

One of the most comprehensive attempts to understand the effects of thermally processing carbohydrate-rich foods on gut microbiota composition, as well as possible mechanisms of action, was a series of interventions conducted in mice (Carmody et al. 2019). First, researchers fed BALB/c mice raw or cooked tuber diets, finding that raw tuber lowered α-diversity and bacterial abundance and raised the relative proportion of Bacteroidetes in comparison with cooked tuber. To test if these results could be due to changes in starch digestibility with cooking, they manipulated starch digestibility in mice diets and found that C57BL/6J mice fed high-digestibility and low-digestibility starch mirrored the patterns in terms of gut microbiota composition of the cooked and raw tuber experiment, respectively. To explore these effects further, they hypothesized that if starch digestibility significantly shaped the gut microbiota, cooking should have a stronger impact on starch-rich foods compared to low-starch foods and, among starch-rich foods, on those with lower ileal digestibility of raw starch. They tested six foods that considered these parameters on C57BL/6J mice, and the results showed significant microbial changes with cooking in foods high in low-digestibility starch (sweet potato and white potato), evidenced by distinct α-diversity and β-diversity patterns. Conversely, low-starch foods (carrot and beet) and high-digestibility starch foods (corn and pea) exhibited no consistent microbial changes from cooking. In summary, this series of experiments suggested that the effects of cooking starch-rich foods on gut microbiota composition could be attributed to changes in starch digestibility.

More research on animals has been conducted to understand the effects of processed carbohydrate-rich foods on gut microbiota composition, generally yielding positive outcomes. Whole chestnut flour and its heat–moisture treated counterpart were incorporated into a high-fat diet given to C57BL/6J mice, and their effects on gut microbiota composition were analyzed (Guo et al. 2024). Unique OTUs were found for the different groups. Nevertheless, α-diversity was not significantly different between the groups. At the genus level, harmful bacteria like *Romboutsia* were reduced after heat–moisture treated whole chestnut flour intervention, while beneficial bacteria such as *Faecalibaculum*, the Lachnospiraceae NK4A136 group, *Desulfovibrio*, *Bifidobacterium*, and *Lachnoclostridium* were increased.

In the same mouse model, the effects of conventional heating and microwaving starch have been compared to those of a normal and a soluble starch diet (Li et al. 2024). The microwave-treated starch group had significantly higher Sobs, ACE, Chao1, and Shannon richness and diversity indexes compared to the normal diet control group. Moreover, more total species were observed in the microwave-treated starch group compared to the other groups. The Firmicutes/Bacteroidetes ratio was lower in the microwave-treated starch group, while *Bacteroides* and *Muribaculum* were significantly higher.

Another investigation was aimed to evaluate the effects of heat–moisture treated high-amylose rice versus its non-heated counterpart on Wistar rats following a high-fat diet, using control and high-fat diets as references (Ma et al. 2023). The Shannon index indicated that α-diversity was higher in the heat–moisture treated group compared to the control with increased levels of beneficial bacteria such as *Bacteroides*, *Parabacteroides*, and *Ruminococcus*.

Researchers have also considered different thermal processing conditions in their experiments. This is the case of a study in which raw, low temperature/long time, high temperature/short time, and ultra-high temperature heated camel milk were injected intraperitoneally to C57BL/6J mice fed a standard chow diet (He et al.

2020). The gut microbiota of mice fed high temperature treated milks clustered closer than that of the mice fed untreated and low temperature/long time treated milk. Furthermore, α-diversity was reduced in response to heat-treated camel milks. Firmicutes were prevalent in the high temperature treated groups, while Bacteroidetes dominated the raw and low temperature/long time groups. At the family level, Muribaculaceae were drastically decreased in mice receiving high temperature treated camel milks compared with mice in the raw and low temperature/long time groups. At the genus level, the abundance of *Bifidobacterium* was lower in the ultra-high temperature group than in the low temperature/long time group. However, in the ultra-high temperature group, *Lactobacillus* was dominant.

Other studies have employed in vitro digestion and fermentation methodologies to conclude that thermal processing benefits gut microbiota composition. One of them analyzed nine types of boiled or steamed whole grains (Fang et al. 2023). Steamed whole grains showed similar microbiota structures, while boiled whole grains displayed greater species variation. At the genus level, *Bifidobacterium* increased especially in the steaming group, while *Megamonas* and *Megasphaera* showed notable increases in steamed barley and boiled oats.

Microwaving β-glucan extracted from whole grain oats has also shown positive effects on gut microbiota composition (Dong et al. 2020). Microwaving β-glucan resulted in higher levels of *Blautia* as well as the highest abundance of *Lactobacillus* and *Bifidobacterium* in comparison to raw and steamed groups and displayed a more significant effect inhibiting *Escherichia–Shigella* compared to the steamed group.

Another study that yielded positive results on gut microbiota composition examined the impact of various processing methods (normal or high-pressure steaming, normal or high-pressure boiling, stir-frying, and microwaving) on the gut microbiota modulatory properties of yam (Gong et al. 2022). Processed yams were generally more effective in promoting beneficial bacteria like *Bifidobacterium* and *Prevotella_9* compared to raw yams. Among all thermal treatments, normal-pressure steaming exhibited the highest prebiotic index value.

However, other studies employing in vitro digestion–fermentation methods showed more favorable effects on gut microbiota composition for unprocessed carbohydrate-rich foods. For instance, one study comparing green and roasted coffee found that green coffee was more effective in increasing the abundance of *Bacteroides* (Pérez-Burillo et al. 2019a). Another investigation that showed results on the same line examined raw, grilled, and fried red-skinned onions (Cattivelli et al. 2023). Raw onions maintained in eubiosis the Firmicutes/Bacteroidetes ratio, while fried onions exceeded the dysbiosis threshold. Regarding beneficial bacterial taxa, raw onions fostered Lactobacillales and *Faecalibacterium prausnitzii*, while cooked onions inhibited beneficial taxa. Regarding opportunistic bacteria, raw onions performed better than fried ones by reducing Enterobacteriaceae, *Clostridium* group I, and the *Atopobium–Collinsella–Eggerthella* group.

Other investigations have shown mixed results. One of them analyzed in vitro digested and fermented grilled, fried, and raw dark purple eggplant (Nissen et al. 2022). Frying dark purple eggplant increased the abundance of Bifidobacteriaceae and Lactobacillales. For Bacteroidetes, raw eggplant exhibited the highest increase, while grilling showed the highest reduction compared to baseline. Raw and fried eggplants maintained eubiosis according to their Firmicutes/Bacteroidetes ratio, while grilling did not. Both *Clostridium* group IV and *Faecalibacterium prausnitzii* significantly increased following fried eggplant fermentation, while *Akkermansia muciniphila* showed the highest increases in raw and fried. Regarding opportunistic bacteria, frying showed the lowest increase for Enterobacteriaceae, while the *Clostridium* group increased significantly for grilling.

Another study offering mixed results aimed to investigate the effects of boiling and stir-frying tomatoes on gut microbiota composition after in vitro digestion and fermentation (Wu et al. 2023). Raw tomatoes had a significantly higher relative abundance of Actinobacteria while both raw and boiled tomatoes reduced the Firmicutes/Bacteroidetes ratio. At the genus level, raw tomatoes exhibited a higher abundance of *Megamonas* and *Bifidobacterium*. *Peptostreptococcus*, associated with colonic inflammation, was significantly lower in the boiled group. *Blautia*, *Lachnospira*, and *Bacteroides* were predominant in raw tomatoes while *Ruminococcus* was the dominant bacterial taxon in the boiled group.

Other results show no impact of thermal food processing on gut microbiota composition. It is the case of unmodified and heat-treated wheat bran submitted to in vitro digestion and individual fermentations with the feces of 10 donors (De Paepe et al. 2019). No significant differences were found between the heat-treated and unmodified bran at the genus and species levels.

2.2.2 Protein-Rich Foods

Regarding the impact of thermal processing on protein-rich foods and its effect on gut microbiota composition, the available literature is limited and the results are inconsistent.

Research has employed animal models to study the effects of thermal food processing on gut microbiota composition. One notable approach involves assessing whether these effects are consistent across various animal models. This is the case of a study that analyzed the effect of two different types of non-thermally and thermally processed fish on the gut microbiota composition of catfish and C57BL/6J mice (Zhang and Li 2018). Both catfish and mice fed thermally processed fish showed significantly lower α-diversity than those fed non-thermally processed fish. Besides the reduction of Bacteroidetes and the balanced Proteobacteria, the response of the gut microbiota to thermally processed food was taxonomically opposite at the phylum and genus levels in both animals. In mice, Bacteroidetes abundance decreased while Firmicutes increased with thermal processing; while in catfish, Fusobacteria increased and Bacteroidetes and Firmicutes decreased. As a result, the Firmicutes/Bacteroidetes ratio was stable or reduced in catfish but increased in mice. Specific genera such as *Cetobacterium* increased in catfish, while *Oscillospira* increased in mice with thermal processing.

Most studies have been conducted only in one animal model. C57BL/6J mice fed AIN-93 M chow or chow containing raw, steamed, baked, or fried oyster showed that the steamed group exhibited the most favorable results for gut microbiota composition (Jiang et al. 2020). The Shannon diversity index decreased in the baked group, while the Simpson index increased in the steamed group compared to baked and fried groups. Steamed oysters increased Firmicutes and decreased Bacteroidetes compared to baked and fried oyster diets. At the family level, unidentified Clostridiales were significantly increased in the baked and fried groups, while Bifidobacteriaceae and Muribaculaceae were predominant in the raw and steamed groups, respectively. At the genus level, *Bifidobacterium*, *Faecalibaculum*, and *Roseburia* were reduced in baked and fried groups but increased in steamed and raw groups. Fried and baked diets enhanced levels of unidentified Clostridiales, *Desulfovibrio*, and *Helicobacter*, while baked oysters led to a significant decrease in Akkermansiaceae and Muribaculaceae. On the other hand, *Lactobacillus* was predominant in the steamed group.

Another investigation on BALB/c mice evaluated changes in their gut microbiota after being fed AIN-93G chow or chow containing 5% or 10% raw or cooked beef loin powder (Kim et al. 2021). This study did not show a clear trend indicating whether the effects of thermal processing on gut microbiota composition were positive or negative. At the phylum level, Firmicutes and Proteobacteria were higher in the raw group, while Bacteroidetes was enriched in the cooked group. Mice fed 10% raw beef loin powder showed a significantly higher relative abundance of Firmicutes and *Lactobacillus* than control mice, and the ratio of Firmicutes/Bacteroidetes showed an increasing trend. Mice fed 10% cooked beef loin powder displayed a higher abundance of Peptostreptococcaceae and a lower abundance of Desulfovibrionaceae compared with the control mice. Regarding genus differences, the cooked group was enriched in *Clostridium*, *Blautia*, *Streptococcus*, *Bacillus*, *Corynebacterium*, *Bacteroides*, and unclassified genera of Bacillaceae and Enterobacteriaceae.

Only one study employed the in vitro digestion–fermentation methodology to focus specifically on the impact of cooking methods in protein-rich foods on gut microbiota composition. Digested and fermented beef, chicken, and salmon processed by frying or boiling were examined (Shen et al. 2010). *Clostridium* spp. were in higher numbers in fried meats compared to boiled meats. Furthermore, *Clostridium histolyticum*/*perfringens* increased particularly in fried meats and more in chicken than beef or salmon. On the other hand, *Coriobacterium*/*Atopobium* showed a significant increase in boiled beef.

2.2.3 Lipid-Rich Foods

Lipid-rich foods, particularly oils processed by deep-frying, have also been studied. The predominant effect on gut microbiota composition was detrimental, though some modest beneficial effects were also observed. In this sense, research has been mainly conducted on animal models. In one experiment, fat extracted from pork belly meat and cooked by braising, stewing, or deep-frying methods was added to animal feed, and the effects of each fat on fecal microbiota composition were analyzed in Sprague Dawley rats compared to rats fed a soybean-oil-treated feed and a blank control (Zhu et al. 2021). The abundance of Bacteroidetes was significantly lower in the cooked fat groups compared to both controls, while an increase in Firmicutes was observed. The proportions of Ruminococcaceae, *Bacteroides*, *Lactobacillus*, *Roseburia*, and *Blautia* varied significantly between groups.

Another investigation to further explore the effects of deep-frying was conducted on Golden Syrian hamsters fed diets containing fresh or deep-fried corn oil or lard (Kwek et al. 2022). Results indicated that deep-fried oil and lard caused the loss of species diversity and richness of the gut microbiota. At the phylum level, the diet with deep-fried corn oil significantly reduced Proteobacteria and Bacteroidetes, and increased Actinobacteria compared to fresh corn oil. The Firmicutes/Bacteroidetes ratio was higher in deep-fried corn oil compared to fresh corn oil. Both deep-fried groups had lower Erysipelotrichaceae and higher Eubacteriaceae compared to both fresh groups. The diet with deep-fried corn oil significantly reduced many families, including Porphyromonadaceae and Lachnospiraceae. At the genus level, both deep-fried groups reduced *Allobaculum* abundance. However, the deep-fried corn oil diet led to increased *Bifidobacterium* compared to fresh corn oil.

Research on the effects of deep-frying other oils has also been performed. The impact of flaxseed oil before and after deep frying on gut microbiota of Sprague Dawley rats was investigated (Zhang et al. 2024). Unprocessed flaxseed oil increased the number of beneficial bacteria, while the relative abundance of obesity-related and inflammatory bacteria was promoted by deep-frying. Deep-fried flaxseed oil treatment increased Negativicutes, Acidaminococcaceae, and Selenomonadales and decreased Lactobacillaceae. Nevertheless, at the species level, an increase in *Lactobacillus* in deep-fried flaxseed oil was observed.

Sunflower and perilla oils have also been studied, showing positive effects of deep-frying on gut microbiota composition. Gut microbiota analysis of Sprague Dawley rats fed high-fat diets supplemented with these oils revealed that the deep-fried perilla oil group was enriched in *Akkermansia*, *Desulfovibrio*, and *Turicibacter*, while the deep-fried sunflower oil group was enriched in *Bifidobacterium*, *Lactobacillus*, and *Breznakia* (Ruan et al. 2023).

Canola oil has also been investigated. Wistar rats fed a basal diet supplemented or not with non-heated or deep-fried canola oil revealed that Bacteroidetes proportions increased in the non-heated canola oil group and decreased in the deep-fried canola oil group (Zhou et al. 2016). While Bacteroidales and Bifidobacteriales were rich in the deep-fried canola oil group, Veillonellaceae and Clostridia were rich in the non-heated canola oil group. At the genus level, the deep-fried canola oil group had the lowest *Prevotella* and the highest *Lactobacillus* proportions while also exhibiting high levels of *Allobaculum*, *Bacteroides*, *Oscillospira*, and *Bifidobacterium*.

Researchers have also compared canola oil to palm oil and their thermally oxidized counterparts, analyzing their effects on the gut microbiota composition of Sprague Dawley rats fed high-fat diets prepared with these oils (Ruan et al. 2021). At the phylum level, the relative abundance of Firmicutes was higher in both palm oil and heated canola oil groups compared to the control, but lower in the non-heated canola oil group. The Firmicutes/Bacteroidetes ratio was the highest in the heated palm oil group followed by the heated canola oil group. The relative abundance of Tenericutes was lower in the heated canola oil group. The heated palm oil group had the highest *Lactobacillus* levels. *Butyrivibrio* and Ruminococcaceae UCG-005 were more abundant in the heated canola oil group, while the non-heated canola oil group had the highest Ruminococcaceae NK4A214 levels. *Roseburia* was higher in the heated palm oil group compared to the non-heated group.

Only one study focused on investigating the effects of thermally processing a lipid-rich food using the methodology of in vitro digestion–fermentation. Gut microbiota analysis of digested and fermented raw and roasted (at four different temperatures) cocoa samples showed a significant dissimilarity (Maldonado-Mateus et al. 2021). The top discriminating bacteria were *Veillonella*, *Faecalibacterium*, and Lachnospiraceae UCG-004. Whereas *Veillonella* was found in much higher abundance in raw samples, *Faecalibacterium*, *Roseburia*, and *Butyricicoccus* were found in higher abundance in roasted samples. The temperature applied did not have an influence on the top seven discriminating genera. Furthermore, negative correlations were found between hydroxymethylfurfural and *Blautia* and *Veillonella,* but a positive significant one was found with *Alistipes*.

2.3 Effect of Food Processing on Gut Microbiota Functionality

2.3.1 Gene Expression

Gene expression analysis can help understand the impact of thermal food processing on gut microbiota functionality by revealing changes in the microbial activity. Analyzing the gene expression of the gut microbiota can be a challenging task, but there are a few available options. Metatranscriptomics is a valuable tool for measuring gene expression of a community of microorganisms. In this sense, using metatranscriptomics, researchers analyzed the impact of feeding BALB/c mice raw or cooked tuber diets on gut microbiota gene expression (Carmody et al. 2019). Community-wide gene expression profiles were clustered by processing method for tuber diets. The microbial communities of mice fed raw tuber diets showed higher expression of genes for the metabolism of starch, sugar and xenobiotic compounds, consistent with the expectations of lower digestibility and higher xenobiotic load in the raw tuber diet. In addition, β-amylase emerged as one of the two enzyme families distinguishing raw versus cooked tuber samples, the other one being a broad glycosyltransferase family.

When metatranscriptomics is not feasible, there are alternatives for the prediction of gut microbiota functionality. Phylogenetic Investigation of Communities by Reconstruction of Unobserved States (PICRUSt) is a bioinformatic tool used to predict the functional composition of a microbial community based on 16S rRNA gene sequencing data. Two studies have used PICRUSt to predict gut microbiota functionality. In one of them, PICRUSt identified five carbohydrate and energy metabolism genes enriched in $ApoE^{-/-}$ mice fed a heat-treated high-fat diet in comparison with mice fed a non-heated high-fat diet (Marungruang et al. 2016). PICRUSt also found an enrichment in genes for glycan biosynthesis and metabolism, including ion-coupled transporters, glycosaminoglycan degradation, and other glycan degradation genes, in BALB/c mice fed AIN-93G chow containing cooked beef loin powder compared to mice fed AIN-93G chow containing raw beef loin powder (Kim et al. 2021).

As a conclusion, it is apparent that the expression of carbohydrate-related genes is key when analyzing the effects of food thermal processing on the functionality of the gut microbiota.

2.3.2 Metabolites

Another approach for examining the impact of thermal food processing on gut microbiota functionality involves analyzing the metabolites produced by microbes. By focusing on these metabolites, researchers can gain insights into how thermal processing alters the gut microbiota's metabolic activities and their subsequent effects on human health.

2.3.2.1 SCFA

One way to explore the effects of food thermal processing on gut microbiota functionality is through the evaluation of SCFA production. Gut microbiota ferment unabsorbed carbohydrates and proteins to SCFA. The production of SCFA by the gut microbiota is an indicator of gut health, as demonstrated by the numerous health benefits they provide to the host. Therefore, analyzing the levels of SCFA produced by gut bacteria in response to different thermal processing methods can give insights into the effects of thermal processing on gut microbiota functionality.

As for the gut microbiota composition studies, SCFA analyses have also been performed comparing thermally processed diets with those that are not thermally processed. It is the case of a study conducted on beagle dogs fed raw, pasteurized, and high-temperature sterilized diets (Cai et al. 2022). The feces of beagles fed raw food showed significantly lower acetic, propionic, and butyric acid levels and total acid content compared to the beagles fed thermally processed diets.

Research comparing the production of SCFA of different foods processed thermally under the same experimental conditions has also been conducted. One study subjected to in vitro digestion and fermentation different processed foods: wheat bread (raw and lightly and highly toasted), chicken (boiled, fried, grilled, and roasted), red pepper (raw, fried, and roasted), chickpeas (boiled and grilled), and banana (raw, roasted, and fried) and studied the SCFA production after fermentation (Pérez-Burillo et al. 2018). For chickpeas, grilling increased butyric acid concentrations significantly, while raw bread produced more butyric acid than toasted bread. For red pepper, roasting increased acetic acid, while frying elevated butyric acid levels. In the case of banana, roasting and frying both enhanced acetic and butyric acid production compared to raw banana. For chicken, roasting, frying, and grilling led to higher acetic and butyric acid levels than boiling.

In general lines, these studies show beneficial effects of thermal processing on the production of SCFA and are very valuable as they study diets or a wide range of food items under the same experimental conditions. Nevertheless, studies that examine the impact of thermal processing on individual food items are essential for gaining a deeper understanding of how thermal processing affects different food matrices.

Carbohydrate-Rich Foods Regarding the impact of thermally processing carbohydrate-rich foods on gut microbiota production of SCFA, a trend of beneficial effects is observed, although some negative impacts or no impacts were also noted.

In animal studies, research has shown beneficial effects. SCFA measured in the feces of C57BL/6J mice fed conventionally heat-treated or microwave-treated starch showed butyric acid was significantly higher in mice fed the microwaved starch diet compared to the conventionally heated starch diet. However, no significant differences between the groups for acetic, propionic, and valeric acids were observed (Li et al. 2024).

Heating camel milk also shows benefits for SCFA production (He et al. 2020). SCFA were measured in the feces of C57BL/6J mice fed a standard chow diet and injected intraperitoneally raw, low temperature/long time, high temperature/short time, or ultra-high temperature camel milk. The SCFA contents were significantly increased in the high-temperature treated camel milk groups compared with the other groups. Notably, the concentration of butyric acid in the high-temperature/short time group was much higher than that in the other groups.

Research using in vitro digestion and fermentation methodology has also demonstrated beneficial effects on SCFA production. It is the case of microwaved β-glucan extracted from whole grain oats, which showed higher concentrations of acetic, propionic, and butyric acids compared to raw and steamed β-glucan after 24 h of fermentation (Dong et al. 2020).

Among the different types of processing methods, some can be more beneficial than others. Depending on the type of whole grain, steaming or boiling may promote greater SCFA production (Fang et al. 2023). Steamed foxtail millet, barley, coix seed, and sorghum showed higher SCFA levels than their boiled counterparts, while oats, quinoa, highland barley, and buckwheat exhibited the opposite trend.

Also, the degree of processing plays a role in SCFA production. One study considering the degree of coffee roasting analyzed the SCFA levels after submitting to in vitro digestion and fermentation of medium- and dark-roasted spent coffee grounds (López-Barrera et al. 2016). Medium-roasted spent coffee grounds produced elevated levels of SCFA after fermentation. In contrast, dark-roasted spent coffee grounds exhibited a higher fermentability index than medium-roasted spent coffee grounds.

Other studies show mixed results regarding the effect of thermal processing on SCFA production. It is the case of a study that aimed to investigate the effects of boiling and stir-frying tomatoes on SCFA production after in vitro digestion and fermentation (Wu et al. 2023). Both raw and boiled tomatoes showed more acetic, propionic, and butyric acid levels than stir-fried ones, although boiled tomatoes produced significantly higher levels of total SCFA than raw and stir-fried samples.

Other research finds no significant impact of thermal processing on gut microbiota SCFA production. It is the case of unheated and dry heat-treated wheat bran subjected to in vitro digestion and fermentation with the feces of 10 healthy donors (De Paepe et al. 2019). The total and individual SCFA production resulting from the fecal microbial fermentation of unmodified and heat-treated wheat bran was not significantly different for the 10 assessed donors.

In contrast, negative impacts of food thermal processing on SCFA production have also been reported. When comparing the production of SCFA of in vitro digested and fermented green and roasted coffee from different origins, green coffee tended to produce higher SCFA amounts, up to 20% more (Pérez-Burillo et al. 2019a). Propionic acid exhibited higher levels in Brazilian and Vietnamese green coffee compared to their roasted coffee counterparts. For butyric acid, Brazilian green coffee showed higher levels compared to its roasted counterpart.

Protein-Rich Foods Not much research has been conducted yet regarding the impact of thermal food processing of protein-rich foods on gut microbiota production of SCFA, and results are mainly not significant or show both beneficial and detrimental effects.

In C57BL/6J mice fed AIN-93 M chow or AIN-93 M chow containing raw, steamed, baked, or fried oyster, all oyster diets promoted acetic and propionic acid contents in feces. However, there were no effects on butyric acid levels except for an increase with steamed oyster (Jiang et al. 2020). The acetic and propionic acid amounts in the baked group were lower than those in the raw, steamed, and fried groups. There were no differences in isobutyric, valeric, and isovaleric acid levels across the oyster groups.

For in vitro digested and fermented beef, chicken, and salmon subjected to frying or boiling, most types of SCFA increased significantly throughout the experiment, but there were few significant differences observed between the cooking methods (Shen et al. 2010). However, after 24 h of fermentation, boiled meat showed higher concentrations than fried meat for *i*-valeric acid.

Lipid-Rich Foods Regarding lipid-rich foods, thermally processing oils had either no impact or a detrimental effect on gut microbiota SCFA production, while thermally processing cocoa had a positive effect.

Golden Syrian hamsters fed diets containing deep-fried corn oil significantly decreased total SCFA by 30% compared to the group fed fresh corn oil (Kwek et al. 2022). This decrease was characterized by reductions in fecal propionic acid (52%), butyric acid (46%), and valeric acid (53%). In the case of hamsters fed diets containing deep-fried or fresh lard, total fecal SCFA levels in both groups were statistically similar.

In another study, feeding Sprague Dawley rats diets containing deep-fried perilla or sunflower oil did not affect SCFA levels (Ruan et al. 2023). Also, no significant differences in acetic and propionic acid levels were observed for rats fed high-fat diets containing thermally oxidized palm and canola oil, although oxidized palm oil resulted in smaller amounts of butyric acid than the control group (Ruan et al. 2021).

However, a positive impact of processing a lipid-rich food on SCFA production has also been reported. In vitro digestion–fermentation of raw and roasted cocoa at four different temperatures showed roasting increased butyric acid production. However, it was also observed that increasing the roasting temperature lowered the levels of acetic and propionic acids (Maldonado-Mateus et al. 2021).

2.3.2.2 (Poly)phenol Metabolites

Food processing can also lead to more efficient interactions of food (poly)phenols with the gut microbiota for the conversion of these compounds into bioactive metabolites. The gut microbiota can also release fiber-bound or entrapped (poly)-phenols within the cell, incrementing their bioaccessibility.

Research in this field suggests that thermal processing enhances both the bioaccessibility and microbial metabolism of (poly)phenols from a quantitative perspective, without affecting the microbial metabolic pathways. However, the specific effects are highly dependent on the type of thermal processing method used. Nevertheless, it should be noted that all available studies were conducted using in vitro digestion and fermentation models, lacking in vivo approaches. As a result, the current information is limited and may not fully capture the complexities of how thermal processing impacts the gut microbiota in a living organism. Further research involving in vivo studies is needed to provide a more comprehensive understanding of these effects.

Pepper is the most studied food regarding the impact of thermal food processing on (poly)phenol microbial metabolism. In the case of green pepper, griddling and frying in olive or sunflower oil significantly impacts the gut microbial metabolism of (poly)phenols (Juániz et al. 2016). Differences in total metabolite amounts were observed between raw and cooked green peppers, but not in the pathways of metabolite formation. Griddled green pepper showed the highest amounts of bioaccessible (poly)phenolic compounds after 24 h of fermentation, followed by sunflower-oil-fried, raw, and olive-oil-fried green peppers. The same order of abundance was followed by the most prevalent phenolic metabolite, 3-(3′-‑hydroxyphenyl)propionic acid.

Another study conducted with peppers analyzed the impact of microwaving, canning, grilling, and frying on the (poly)phenol metabolites of Piquillo peppers (Burgo-Gutiérrez et al. 2024). Similar phenolic derivatives were detected in raw and heat-treated samples after fermentation, although concentration and time of appearance were different among samples. After 24 h of fermentation, the concentration of total (poly)phenolic compounds was the highest in fried, followed by microwaved, grilled, canned, and raw Piquillo peppers. The most abundant phenolic derivative generated at 24 h was 3-(3′-hydroxyphenyl)propionic acid, which exhibited more than twofold higher amounts in heat-treated samples compared to raw ones, with frying being the method that exhibited its highest abundance, followed by microwaving, canning, grilling, and raw.

The impact of thermal processing on the (poly)phenol microbial metabolites of dark purple eggplant has also been studied (Nissen et al. 2022). Researchers found that the metabolic pathways involved in the metabolization of eggplant (poly)-phenols were consistent across raw, fried, and grilled samples, with cooking primarily affecting the quantity of metabolites rather than the pathways themselves. Frying significantly enhanced the microbial metabolism of phenolic compounds from eggplant, while metabolites were also produced in grilled and raw samples but in lower quantities. After 24 h of fermentation, the primary metabolite remaining was 3-(3′-hydroxyphenyl)propionic acid, which accounted for 89%, 93%, and 95% of total phenolics in raw, fried, and grilled samples, respectively.

Grilling and frying red onion also significantly impacted (poly)phenol metabolism (Cattivelli et al. 2023). The highest concentration of (poly)phenol metabolites was found in grilled onions, followed by fried and raw onions. 3-(3′-hydroxyphenyl)propionic acid, the predominant metabolite across all samples, was the highest in grilled onions, followed by fried and raw ones. Other significant metabolites included 3-(3′-hydroxyphenyl)acetic acid, 3,4-dihydroxybenzoic acid, and 3-(3′,4′-dihydroxyphenyl)acetic acid, with their proportions varying depending on the thermal treatment. Moreover, in fried onions, the metabolite 3-(3′,4′-dihydroxyphenyl)acetic acid was found in high proportion, whereas it represented only 2.7% and 4.3% of the total metabolites in grilled and raw onion samples, respectively.

The effect of thermal processing on the phenolic metabolites of cardoon has also been analyzed (Juániz et al. 2017). Grilling and frying cardoon with olive or sunflower oil primarily affected the total amount of metabolites produced rather than the metabolic pathways of metabolite production. After 24 h of fermentation, griddled cardoon showed the highest amount of bioaccessible (poly)phenols, followed by cardoon fried in sunflower oil, cardoon fried in olive oil, and raw cardoon. The same order of abundance was followed by the principal metabolite, 3-(3′-hydroxyphenyl)propionic acid.

In the case of artichokes, boiling, sous vide, and microwaving treatments had a significant impact on the total metabolites and (poly)phenols measured after 24 h of fermentation, which were higher in the sous vide artichokes followed by the boiled, microwaved, and raw artichokes (Domínguez-Fernández et al. 2021).

In the case of coffee, roasting has been shown to increase the amounts of caffeic and ferulic acids as well as the phenolic metabolites *m*-hydroxybenzoic acid and *p*-hydroxyphenylacetic acid in comparison with green coffee after in vitro fermentation (Pérez-Burillo et al. 2019a).

2.3.3 Other Functionalities

Evaluating gene expression and microbial metabolites is the best approach to understand changes in gut microbiota functionality due to thermal processing.

However, the impact of food thermal processing on gut microbiota functionality can also be assessed by evaluating other specific measurements after incubating the processed food with gut microbiota. In this category, we include the effects of food thermal processing on the bioaccessibility of mineral elements and chemical compounds as well as the changes in the antioxidant capacity of foods measured after fermentation with gut microbiota. However, it should be noted that this approach does not involve measuring a direct outcome of gut microbiota action. Instead, it measures a variable after a process in which the gut microbiota participates, attributing the observed changes to its activity.

Regarding impacts on bioaccessibility, research has been conducted on minerals and carotenoids. In the case of carotenoids, one study aimed to investigate the effects of boiled and stir-fried tomatoes compared to raw ones on the carotenoid bioaccessibility after in vitro digestion and fermentation (Wu et al. 2023). After fermentation, lycopene and β-carotene levels were higher in raw and boiled samples than in stir-fried samples, which could indicate that the gut microbiota is able to extract more carotenoids from raw or lightly cooked foods.

In the case of minerals, research has been conducted for zinc and selenium. In one study, 55 different foods from animal and vegetal origin processed by frying, boiling, roasting, grilling, and toasting were subjected to in vitro digestion and fermentation with feces from healthy adults and children. Zinc concentrations were measured after fermentation (García-Conde et al. 2023). When considering all foods or dividing them into plant and animal foods, there were no significant differences among cooking methods for the zinc bioaccessibility in the adults' fermentations. However, after children's fermentations, when considering all foods together or dividing them according to their origin, significant differences depending on the processing method used were observed. Frying was related to the highest zinc bioaccessibility values in all foods, plant- or animal-based. Zinc bioaccessibility was significantly lower for raw foods compared to those submitted to frying, roasting, and boiling. On the same samples, researchers assessed selenium bioaccessibility (García-Conde et al. 2024). After fermentation, significantly lower values for the raw vegetables were observed compared to the roasted, boiled, and grilled vegetables. The same behavior was observed for animal foods, with the exception of grilling. When comparing the results of the adults' and children's fermentations, selenium bioaccessibility of fried, roasted, and boiled plant foods was significantly lower after fermentation with feces of adults than with those of healthy children.

Taking this information altogether, for carotenoids, raw and mild cooking techniques like boiling promoted their bioaccessibility after fermentation, while in the case of minerals like zinc and selenium, bioaccessibility was improved by stronger processing methods like frying or roasting, although the effects varied depending on whether the fermentations were carried out using feces from adults or children.

Beyond its impact on the bioaccessibility of elements and compounds, the gut microbiota can positively contribute to the antioxidant capacity of foods by producing or releasing antioxidants from the food matrix. Conversely, it can also have a detrimental effect by metabolizing antioxidants. Thermal processing plays a crucial role by modifying foods and producing, degrading, or releasing antioxidant molecules. Therefore, it is essential to understand how the gut microbiota influences the release of antioxidant capacity of foods having undergone different thermal processes.

This topic has not been extensively studied, and the methodologies employed are primarily based on in vitro digestion and fermentation models. The few available studies indicate a significant dependence on the thermal processing method used and the assay employed for measuring antioxidant capacity, with a tendency toward negative or nonsignificant impacts on antioxidant capacity.

Both animal and plant foods have been studied. In the case of animal foods, 11 different foods processed by frying, boiling, roasting, and grilling showed no differences between processed and raw foods in assays measuring antioxidant capacity against 2,2-diphenyl-1-picrylhydrazyl radicals (DPPH) or in those measuring the Ferric Reducing Antioxidant Power (FRAP). However, foods treated with ultra-high temperature exhibited a significantly lower antioxidant capacity than raw foods in the FRAP assay (Navajas-Porras et al. 2021).

In the case of plant foods, 42 different plant foods were processed by frying, boiling, roasting, grilling, brewing, or toasting, and the antioxidant capacity after in vitro digestion–fermentation was analyzed using DPPH, FRAP, and Folin–Ciocalteu assays (Navajas-Porras et al. 2020). In the case of the DPPH assay, the antioxidant capacity of the fermented fraction was significantly lower in boiled and toasted vegetables compared to raw foods. On the other hand, the Folin–Ciocalteu assay showed a significantly lower antioxidant capacity in brews and toasted foods compared to raw foods. The FRAP results were on the same line, showing a significantly lower antioxidant capacity in brews compared to raw foods.

In the case of cocoa roasted at four different temperatures, a higher degree of roasting decreased antioxidant capacity for DPPH, FRAP, and Folin–Ciocalteu assays after fermentation (Maldonado-Mateus et al. 2021).

Research also demonstrated that the antioxidant capacity measured by DPPH, FRAP, and Folin–Ciocalteu assays after fermentation of 20 foods from animal and vegetal origin processed by frying, boiling, roasting, grilling, and toasting showed almost no significant correlations with heat damage markers (furosine, hydroxymethylfurfural, and furfural). The exception was roasted foods, for which positive significant correlations were found between furosine and the antioxidant capacity measured by Folin–Ciocalteu and FRAP assays (Navajas-Porras et al. 2022).

Other experiments exhibit mixed results. It is the case of digested and fermented green and roasted coffee from three different locations (Pérez-Burillo et al. 2019a). The 2,2-azino-bis-3-ethylbenzothiazoline-6-sulfonic acid (ABTS) assay showed fermented coffee brews did not exhibit significant differences, except for Brazilian green coffee, which showed higher antioxidant capacity than its roasted counterpart. For the FRAP assay, fermented samples showed no significant differences between green and roasted coffee for any coffee type. For the Folin–Ciocalteu assay, the only one that displayed significant differences was the roasted Vietnamese compared to its green counterpart. A different study showed an increase of antioxidant capacity after fermentation with gut microbiota associated with cooking methods. In this study, 23 different vegetables underwent various cooking methods, including steaming, boiling, grilling, roasting, frying, and breading, with both standard and extended processing times (Pérez-Burillo et al. 2019b). After fermentation, antioxidant capacities were measured against ABTS, FRAP, and OH radicals indicating that frying, breading, and grilling could enhance the antioxidant capacity of vegetables, although the impact varied depending on the antioxidant capacity assay used. Also, both standard and extended cooking times improved antioxidant capacity compared to raw vegetables.

2.4 Effect of Compounds Generated by Food Processing on Gut Microbiota Composition and Functionality

Certain compounds are produced during thermal food processing and could be behind the effects of thermal food processing on gut microbiota composition and functionality. However, the effects of these compounds should not be extrapolated to the effects of thermal food processing, as studying the effect of these compounds outside a food matrix can lead to different results than studying the processed food itself. Nevertheless, it is a research field that is being explored and that could aid in understanding the effects of food processing on gut microbiota composition and functionality.

Among all the products generated during thermal food processing, the main compounds studied are Maillard reaction products, followed by acrylamide and heterocyclic amines. The relationship of these compounds with the gut microbiota is dual: these compounds influence the composition and functionality of the gut microbiota, and in turn, the gut microbiota can metabolize these compounds.

Regarding the metabolization of these compounds by the gut microbiota, metabolic pathways for the degradation of heterocyclic amines have been proposed (Beer et al. 2019; Zhang et al. 2019). For Maillard reaction products, the capacity of the human colonic microbiota to degrade them has also been demonstrated (Hellwig et al. 2015; Yuan et al. 2023).

Research on the impact of these compounds on gut microbiota composition and functionality is inconclusive regarding whether the effects are positive or negative. Some compounds show evidence of detrimental effects, while others demonstrate more beneficial impacts, indicating that the effects are highly dependent on the specific compound studied.

Acrylamide is a compound that has demonstrated detrimental effects, with studies indicating that it negatively impacts gut microbiota composition and functionality. However, research on the specific effects of acrylamide on the gut microbiota remains limited (Yan et al. 2023).

On the other hand, there is significantly more research on Maillard reaction products, though the results remain contradictory. Among Maillard reaction products, melanoidins show the best effects on gut microbiota composition and functionality. Melanoidins are formed at late condensation stages of the Maillard reaction and are found in a multicomponent polymer complex incorporating proteins, carbohydrates, (poly)phenols, and minerals. Melanoidins evade gastrointestinal digestion to get fermented in the colon by the gut microbiota, exhibiting prebiotic effects by promoting beneficial bacteria and inhibiting the growth of pathogenic bacteria. Melanoidin fermentation also increases SCFA production and enhances the antioxidant capacity released after fermentation (Shaheen et al. 2022).

Another significant group of Maillard reaction products is advanced glycation end products. The research about the effects of these compounds on gut microbiota composition and functionality is rather inconclusive. Human and animal trials demonstrate that they are able to modulate gut microbiota composition, with some boosting beneficial bacteria, while others promote detrimental bacteria (Pérez-Burillo et al. 2017). These discrepancies could be attributed to the fact that advanced glycation end products comprise a very complex group of compounds. The number of studies targeting the gut microbiota is still limited and the mechanisms explaining the current findings are not clear (Miclotte and Van de Wiele 2020). However, it has been observed that the effect of advanced glycation end products on the gut microbiota is dependent on the type of species, heating method, and exposure dose. Also, free advanced glycation end products, such as carboxymethyllysine, have detrimental effects on the composition and functions of the gut microbiota, whereas bound advanced glycation end products have a more beneficial effect, although certain detrimental effects may be also observed (Yuan et al. 2023; Aschner et al. 2023). At present, studies on whether free advanced glycation end products affect microbial metabolites are lacking, and the effect of bound advanced glycation end products on the production of SCFA remains controversial (Yuan et al. 2023).

2.5 Conclusions

Research on the effects of food thermal processing on gut microbiota composition and functionality is still limited, although there has been a notable increase in studies in the last years, as evidenced by the recentness of the cited literature. The scarcity of studies should be addressed given the widespread consumption of thermally processed foods in modern societies and the significant impact of gut microbiota on health.

The available research yielded mixed results regarding whether thermal food processing affects gut microbiota composition and functionality. The outcomes depended on the type of food, processing method, and measurement, making it difficult to draw a definitive conclusion on whether the overall effects are positive or negative.

Regarding the impact on gut microbiota composition, this ambiguity partly arises from the inherent challenges of gut microbiota composition research. Defining a "healthy" microbiota and determining the most relevant metrics to measure it, such as an increase in beneficial bacteria, a reduction in detrimental bacteria, or a low Firmicutes/Bacteroidetes ratio, are challenges that complicate the assessment of thermal processing effects. Studies often report data at different specific taxonomic levels, further complicating the extraction of overall conclusions. Despite these common obstacles in microbiota research, some trends have been observed. The effect of thermal processing is very dependent on the type of food and processing method assessed. Thermal processing of carbohydrate-rich foods tends to show the most favorable effects, while thermally processed lipid-rich foods are the least beneficial. Protein-rich foods are less studied in this context, resulting in mixed outcomes. Microwaving is one processing method that has shown positive effects on gut microbiota composition. Additionally, the degree of thermal processing impacts gut microbiota composition, and responses vary depending on the animal model used in the studies.

Regarding the effect of food thermal processing on gut microbiota functionality, the outcomes depend on the specific aspect of functionality assessed, as well as the type of food and processing method studied. For gene expression of the gut microbiota, more research is needed, but the analyses conducted so far suggest a significant role for carbohydrate metabolism genes. For SCFA, the results mirrored gut microbiota composition findings: thermal processing of carbohydrate-rich foods was more beneficial while foods rich in proteins showed inconclusive results due to the small number of studies, and thermal processing of lipid-rich foods was more detrimental to SCFA production. For (poly)phenols, there were clearly positive effects of thermal processing on both (poly)phenol bioaccessibility and microbial metabolite production, with no impact on the metabolic pathways. Regarding the assessments of mineral and carotenoid bioaccessibility, and antioxidant capacity, positive and negative effects were, respectively, observed. However, it is important to note that these studies provide an indication of potential microbiota activity but are not direct measurements of it. Across the different functionalities, microwaving was one of the processing methods that exhibited the most beneficial results. Finally, it was also observed that the degree of thermal processing impacted functionality measurements.

Regarding the effects of compounds generated during thermal processing on gut microbiota composition and functionality, the results are inconclusive and highly dependent on the specific compound. Positive effects have been observed for melanoidins, while compounds like acrylamide have shown negative effects. Additionally, it has been demonstrated that the gut microbiota can metabolize some of these compounds.

Nevertheless, we want to emphasize the heterogeneity in the results. This variability is present within studies, where different measurements may contradict each other, and between studies, where composition and functionality assessments often diverge. Some studies conclusively identify one thermal processing method as the best across various measurements, but making a general conclusion from all studies is challenging due to this intrastudy and interstudy heterogeneity. This inconsistency could arise from the different thermal food processing parameters considered in the studies, with temperature and duration of heat being the most critical variables to control. Additionally, factors such as food size, cooking utensils, and type of oil used in frying may also contribute to the observed heterogeneity.

It is also important to interpret these results with caution due to the limitations of the studies. Regarding the study design, only one human intervention study was performed. In vitro digestion–fermentation experiments are the most prevalent, followed by in vivo experiments in various animal models, which may offer only a limited understanding of what to expect in humans.

Regarding the measurements, some studies focus solely on either gut microbiota composition or a specific aspect of functionality. For a more comprehensive understanding, studies should consider both composition and functionality measurements. It would also be beneficial to correlate composition with functionality as well as with markers of thermal processing to seek possible explanations and help elucidate the underlying mechanisms involved. To comprehensively understand the mechanisms through which thermal food processing affects gut microbiota composition and functionality, it is essential to integrate various omics sciences. While microbial genomics and metabolomics are widely used, other omics should be explored. Metatranscriptomics, which could provide significant insights into functionality, has been only scarcely used.

The effects of thermal food processing on gut microbiota can vary significantly based on the composition of an individual's microbiota, making this an important aspect to include in research. Personalized nutrition could take this into account, as food processing is a modifiable aspect of the diet. This approach could help tailor food processing methods to better suit each individual's gut microbiota, optimizing its composition and functionality on a personalized level.

Future research in this field should explore a broader variety of food types and processing methods, accounting for their parameters, while also incorporating new measurements of microbiota composition and functionality. Human studies should be conducted to gain a clearer understanding of the actual effects. Additionally, longitudinal studies are needed to understand the long-term effects of the frequent consumption of thermally processed foods in our diets. Lastly, taking into account individual gut microbiota responses to thermally processed foods would offer an opportunity for advancements in personalized nutrition.

Acknowledgments This work is part of the thesis of Adriana Delgado-Osorio for the PhD program in Nutrition and Food Science at the University of Granada and is supported by a Formación de Profesorado Universitario predoctoral fellowship (FPU20/03838) granted by the Spanish Ministry of Science, Innovation and Universities.

Competing Interests The authors declare that they have no competing interests.

References

Aschner M, Skalny AV, Gritsenko VA, Kartashova OL, Santamaria A, Rocha JBT et al (2023) Role of gut microbiota in the modulation of the health effects of advanced glycation end-products (review). Int J Mol Med 51(5):1–12

Ashkar F, Wu J (2023) Effects of food factors and processing on protein digestibility and gut microbiota. J Agric Food Chem 71(23):8685–8698

Beer F, Urbat F, Franz CMAP, Huch M, Kulling SE, Bunzel M et al (2019) The human fecal microbiota metabolizes foodborne heterocyclic aromatic amines by Reuterin conjugation and further transformations. Mol Nutr Food Res 63(10):1801177

Burgo-Gutiérrez CD, Ludwig IA, Peña MPD, Cid C (2024) Industrial and culinary treatments applied to Piquillo pepper (Capsicum annuum cv. Piquillo) impact positively on (poly)phenols' bioaccessibility and gut microbiota catabolism. Food Funct 15(5):2443–2458

Cai X, Liao R, Chen G, Lu Y, Zhao Y, Chen Y (2022) The influence of food processing methods on serum parameters, apparent total-tract macronutrient digestibility, fecal microbiota and SCFA content in adult beagles. PLoS One 17(1):e0262284

Carmody RN, Bisanz JE, Bowen BP, Maurice CF, Lyalina S, Louie KB et al (2019) Cooking shapes the structure and function of the gut microbiome. Nat Microbiol 4(12):2052–2063

Cattivelli A, Nissen L, Casciano F, Tagliazucchi D, Gianotti A (2023) Impact of cooking methods of red-skinned onion on metabolic transformation of phenolic compounds and gut microbiota changes. Food Funct 14(8):3509–3525

De Paepe K, Verspreet J, Courtin CM, Van de Wiele T (2019) Wheat bran thermal treatment in a hot air oven does not affect the fermentation and colonisation process by human faecal microbiota. J Funct Foods 60:103440

Domínguez-Fernández M, Ludwig IA, Peña MPD, Cid C (2021) Bioaccessibility of Tudela artichoke (Cynara scolymus cv. Blanca de Tudela) (poly)phenols: the effects of heat treatment, simulated gastrointestinal digestion and human colonic microbiota. Food Funct 12(5): 1996–2011

Dong J-l, Yang M, Zhu Y-y, Shen R-l, Zhang K-y (2020) Comparative study of thermal processing on the physicochemical properties and prebiotic effects of the oat β-glucan by *in vitro* human fecal microbiota fermentation. Food Res Int 138:109818

Fang F, He Y, Zhao J, Zhang Y, Chen C, He H et al (2023) Effects of boiling and steaming process on dietary fiber components and in vitro fermentation characteristics of 9 kinds of whole grains. Food Res Int 164:112328

García-Conde Ú, Navarro-Alarcón M, Navajas-Porras B, Hinojosa-Nogueira D, Delgado-Osorio A, Pérez-Burillo S et al (2023) Total Zn of foods and bioaccesible fractions in the small and large intestine after in vitro digestion and fermentation with fecal material of healthy adults and children: influence of culinary techniques. Food Res Int 169:112817

García-Conde Ú, Navarro-Alarcón M, Navajas-Porras B, Hinojosa-Nogueira D, Delgado-Osorio A, Navarro-Moreno M et al (2024) Selenium bioaccesibility after *in vitro* digestion/fermentation of foods differs in adults and children. Food Biosci 59:103964

Gong L, Hu L, Liu F, Chi J, Chen R, Wang J (2022) Domestic cooking affects the prebiotic performances of Chinese Yam. Food Secur 11(23):3794

Guo Q, Chen L, Yang D, Zheng B (2024) Heat-moisture treatment enhances the ordered degree of starch structure in whole chestnut flour and alters its gut microbiota modulation in mice fed with high-fat diet. Int J Biol Macromol 254:127961

He J, Sun R, Hao X, Battulga A, Juramt N, Yi L et al (2020) The gut microbiota and its metabolites in mice are affected by high heat treatment of Bactrian camel milk. J Dairy Sci 103(12): 11178–11189

Hellwig M, Bunzel D, Huch M, Franz CMAP, Kulling SE, Henle T (2015) Stability of individual Maillard reaction products in the presence of the human colonic microbiota. J Agric Food Chem 63(30):6723–6730

Jiang S, Zeng M, Zhao Y (2020) Thermal processed *Crassostrea gigas* impact the mouse gut microbiota. J Funct Foods 75:104254

Juániz I, Ludwig IA, Bresciani L, Dall'Asta M, Mena P, Del Rio D et al (2016) Catabolism of raw and cooked green pepper (*Capsicum annuum*) (poly)phenolic compounds after simulated gastrointestinal digestion and faecal fermentation. J Funct Foods 27:201–213

Juániz I, Ludwig IA, Bresciani L, Dall'Asta M, Mena P, Del Rio D et al (2017) Bioaccessibility of (poly)phenolic compounds of raw and cooked cardoon (*Cynara cardunculus* L.) after simulated

gastrointestinal digestion and fermentation by human colonic microbiota. J Funct Foods 32: 195–207
Kim HJ, Kim D, Kim KW, Lee SH, Jang A (2021) Comparative analysis of the gut microbiota of mice fed a diet supplemented with raw and cooked beef loin powder. Sci Rep 11(1):11489
Kwek E, Yan C, Ding H, Hao W, He Z, Ma KY et al (2022) Effects of thermally-oxidized frying oils (corn oil and lard) on gut microbiota in hamsters. Antioxidants 11(9):1732
Lerma-Aguilera AM, Pérez-Burillo S, Navajas-Porras B, León ED, Ruíz-Pérez S, Pastoriza S et al (2024) Effects of different foods and cooking methods on the gut microbiota: an in vitro approach. Front Microbiol 14:1334623
Li N, Cheng Z, Qiao D, Li K, Zhao S, Zhang B (2024) Ingesting microwave-cooked starch could tailor the intestinal microbiota and glucose/lipid level of mice fed with low-fat diet. ACS Food Sci Technol 4(2):355–364
López-Barrera DM, Vázquez-Sánchez K, MaGF L-P, Campos-Vega R (2016) Spent coffee grounds, an innovative source of colonic fermentable compounds, inhibit inflammatory mediators in vitro. Food Chem 212:282–290
Ma S, Takasugi S, Sugawara M, Saito K, Jia H, Kato H (2023) Effects of heat-moisture-treated high-amylose Rice flour on body weight, lipid metabolism, and gut microbiome composition in obese rats. Meta 13(7):858
Maldonado-Mateus LY, Perez-Burillo S, Lerma-Aguilera A, Hinojosa-Nogueira D, Ruíz-Pérez S, Gosalbes MJ et al (2021) Effect of roasting conditions on cocoa bioactivity and gut microbiota modulation. Food Funct 12(20):9680–9692
Marungruang N, Fåk F, Tareke E (2016) Heat-treated high-fat diet modifies gut microbiota and metabolic markers in apoe−/− mice. Nutr Metab 13(1):22
Miclotte L, Van de Wiele T (2020) Food processing, gut microbiota and the globesity problem. Crit Rev Food Sci Nutr 60(11):1769–1782
Navajas-Porras B, Pérez-Burillo S, Valverde-Moya ÁJ, Hinojosa-Nogueira D, Pastoriza S, Rufián-Henares JÁ (2020) Effect of cooking methods on the antioxidant capacity of plant foods submitted to in vitro digestion–fermentation. Antioxidants. 9(12):1312
Navajas-Porras B, Pérez-Burillo S, Valverde-Moya Á, Hinojosa-Nogueira D, Pastoriza S, Rufián-Henares JÁ (2021) Effect of cooking methods on the antioxidant capacity of foods of animal origin submitted to in vitro digestion-fermentation. Antioxidants. 10(3):445
Navajas-Porras B, Pérez-Burillo S, Hinojosa-Nogueira D, Pastoriza S, Rufián-Henares JÁ (2022) Relationship of thermal treatment and antioxidant capacity in cooked foods. Antioxidants 11(12):2324
Neilson AP, Ferruzzi MG (2011) Influence of formulation and processing on absorption and metabolism of Flavan-3-Ols from tea and cocoa. Annu Rev Food Sci Technol 2:125–151
Nissen L, Cattivelli A, Casciano F, Gianotti A, Tagliazucchi D (2022) Roasting and frying modulate the phenolic profile of dark purple eggplant and differently change the colon microbiota and phenolic metabolites after in vitro digestion and fermentation in a gut model. Food Res Int 160:111702
Pérez-Burillo S, Pastoriza S, Rufián-Henares JA, Delgado-Andrade C (2017) Effects of dietary AGEs in the gut microbiota composition. In: Dietary AGEs and their role in health and disease. CRC Press
Pérez-Burillo S, Pastoriza S, Jiménez-Hernández N, D'Auria G, Francino MP, Rufián-Henares JA (2018) Effect of food thermal processing on the composition of the gut microbiota. J Agric Food Chem 66(43):11500–11509
Pérez-Burillo S, Mehta T, Esteban-Muñoz A, Pastoriza S, Paliy O, Ángel R-HJ (2019a) Effect of in vitro digestion-fermentation on green and roasted coffee bioactivity: the role of the gut microbiota. Food Chem 279:252–259
Pérez-Burillo S, Rufián-Henares JÁ, Pastoriza S (2019b) Effect of home cooking on the antioxidant capacity of vegetables: relationship with Maillard reaction indicators. Food Res Int 121:514–523

Rajakaruna S, Pérez-Burillo S, Rufián-Henares JÁ, Paliy O (2024) Human gut microbiota fermentation of cooked eggplant, garlic, and onion supports distinct microbial communities. Food Funct 15(5):2751–2759

Rinninella E, Raoul P, Cintoni M, Franceschi F, Miggiano G, Gasbarrini A et al (2019) What is the healthy gut microbiota composition? A changing ecosystem across age, environment, diet, and diseases. Microorganisms 7(1):14

Ruan M, Bu Y, Wu F, Zhang S, Chen R, Li N et al (2021) Chronic consumption of thermally processed palm oil or canola oil modified gut microflora of rats. Food Sci Human Wellness 10(1):94–102

Ruan M, Zhang Z, Yuan X, Zhou R, Zhang S, Tian Y et al (2023) Effects of deep frying vegetable oils rich in PUFAs on gut microbiota in rats. Int J Food Sci Technol 58(1):37–44

Shaheen S, Shorbagi M, Lorenzo JM, Farag MA (2022) Dissecting dietary melanoidins: formation mechanisms, gut interactions and functional properties. Crit Rev Food Sci Nutr 62(32): 8954–8971

Shen Q, Chen YA, Tuohy KM (2010) A comparative *in vitro* investigation into the effects of cooked meats on the human faecal microbiota. Anaerobe 16(6):572–577

Tannock GW (2023) Understanding the gut microbiota by considering human evolution: a story of fire, cereals, cooking, molecular ingenuity, and functional cooperation. Microbiol Mol Biol Rev 88(1):e00127–e00122

Tomás-Barberán FA, Espín JC (2019) Effect of food structure and processing on (poly)phenol–gut microbiota interactions and the effects on human health. Annu Rev Food Sci Technol 10:221–238

Tuohy KM, Hinton DJS, Davies SJ, Crabbe MJC, Gibson GR, Ames JM (2006) Metabolism of Maillard reaction products by the human gut microbiota—implications for health. Mol Nutr Food Res 50(9):847–857

van Boekel M, Fogliano V, Pellegrini N, Stanton C, Scholz G, Lalljie S et al (2010) A review on the beneficial aspects of food processing. Mol Nutr Food Res 54(9):1215–1247

Wu S, Bhat ZF, Gounder RS, Mohamed Ahmed IA, Al-Juhaimi FY, Ding Y et al (2022) Effect of dietary protein and processing on gut microbiota—a systematic review. Nutrients 14(3):453

Wu X, Zhu C, Zhang M, Wang S, Yu J, Tian J et al (2023) Effects of different processed tomatoes on carotenoid release and microbiota composition during in vitro gastrointestinal digestion and colonic fermentation. Food Funct 14(22):10177–10187

Yan F, Wang L, Zhao L, Wang C, Lu Q, Liu R (2023) Acrylamide in food: occurrence, metabolism, molecular toxicity mechanism and detoxification by phytochemicals. Food Chem Toxicol 175: 113696

Yuan X, Nie C, Liu H, Ma Q, Peng B, Zhang M et al (2023) Comparison of metabolic fate, target organs, and microbiota interactions of free and bound dietary advanced glycation end products. Crit Rev Food Sci Nutr 63(19):3612–3633

Zhang Z, Li D (2018) Thermal processing of food reduces gut microbiota diversity of the host and triggers adaptation of the microbiota: evidence from two vertebrates. Microbiome 6(1):99

Zhang J, Lacroix C, Wortmann E, Ruscheweyh HJ, Sunagawa S, Sturla SJ et al (2019) Gut microbial beta-glucuronidase and glycerol/diol dehydratase activity contribute to dietary heterocyclic amine biotransformation. BMC Microbiol 19(1):99

Zhang X, Chen J, Zhou S, Jiang Y, Wang Y, Li Y (2024) The effect of flaxseed oil after deep frying on lipid metabolism and gut barrier homeostasis. Food Res Int 175:113728

Zhou Z, Wang Y, Jiang Y, Diao Y, Strappe P, Prenzler P et al (2016) Deep-fried oil consumption in rats impairs glycerolipid metabolism, gut histology and microbiota structure. Lipids Health Dis 15(1):86

Zhu W, Xu Y, Liu J, Chen D, Zhang H, Yang Z et al (2021) Effects of dietary pork fat cooked using different methods on glucose and lipid metabolism, liver inflammation and gut microbiota in rats. Food Secur 10(12):3030

Recycling Plant-Based Foods for Microbiome-Targeted Personalized Nutrition

3

Ester Betoret, Noelia Betoret, and María José Gosalbes

Abstract

Currently, the European Commission through the Green Pact (the Green Deal) has indicated that food production must be sustainable considering climate change, avoiding the loss of biodiversity, and minimizing residues and food loss/waste. To address these challenges, the 2030 Agenda promotes research on healthy, accessible, and sustainable food, in which consideration of food effects on the human microbiome plays a crucial role due to its close relationship with human health.

Food loss is one of the biggest problems faced by the current food system with negative economic and nutritional impacts and a large environmental footprint. Since fruit, vegetable, and nut by-products are good natural sources of macronutrients (carbohydrates, polysaccharides, proteins, fats, and fiber) and micronutrients (vitamins, minerals, and antioxidants), different valorization strategies have been applied. Among the various processes for stabilizing food losses, dehydration treatment combined with milling is a sustainable alternative for producing functional powdered ingredients with prebiotic effects on the gut microbiota. In this way, a wide range of prebiotics from different origins are being developed to improve human health through the modulation of the microbiome.

E. Betoret · N. Betoret
Instituto Universitario de Ingeniería de Alimentos –FoodUPV, Universitat Politècnica de València, Valencia, Spain

M. J. Gosalbes (✉)
Fundación para el Fomento de la Investigación Sanitaria y Biomédica de la Comunidad Valenciana-Salud Pública, Valencia, Spain

CIBER de Epidemiología y Salud Pública, Instituto de Salud Carlos III, Madrid, Spain
e-mail: Maria.Jose.Gosalbes@uv.es

M. P. Francino, J. Á. Rufián Henares (eds.), *The Microbiome and Personalized Nutrition*, The Microbiomes of Humans, Animals, Plants, and the Environment 6,
https://doi.org/10.1007/978-3-031-93902-0_3

Keywords

Food by-products · Bioactive compounds · Prebiotics · In vitro digestion · Colonic fermentation · Ggut microbiome

3.1 Agricultural Waste and By-Products

3.1.1 Food Losses: Types and Issues

Sustainable Development Goal (SDG) 12, adopted in September 2015, aims to "ensure sustainable consumption and production patterns." Included in it, Target 12.3 aims to "halve per capita global food waste at the retail and consumer level by 2030 and reduce food losses along the food production and supply chains" (https://sdgs.un.org/goals, 2024). In the European Union, Directive 2008/98/EC defined a waste hierarchy that must apply as a priority order in waste prevention and management legislation and policy in the member states. The hierarchy includes "(1) prevention, (2) preparing for reuse, (3) recycling, (4) other recovery (e.g., energy recovery), and (5) disposal." Similarly, according to the US Environmental Protection Agency, the Food Recovery Hierarchy ranks the most important steps to prevent and redirect food waste including "(1) source reduction, (2) feed hungry people, (3) feed animals, (4) industrial uses, (5) composting, and (6) landfill/incineration" (EPA 2022).

The terms "food loss" and "food waste" are commonly mixed in the literature. Food and Agriculture Organization of the United Nations (FAO) has proposed clarification of the concepts, so food loss refers to losses that occur "along the food supply chain from harvest up to, but not including, the retail level," and food waste occurs "at the retail and consumption level" (FAO 2019). Given the importance of reducing food losses and waste, FAO and the United Nations Environment Programme are now measuring the Food Loss Index and Food Waste Index to evaluate the progress toward Target 12.3. The first report from FAO in 2019 showed that 14% of food produced is lost from postharvest up to, but excluding, the retail level (FAO 2019). However, FAO estimated that this percentage is much higher given the difficulty of measuring all the losses (FAO 2019). In 2021, the percentage of food lost globally after harvest on farm, transport, storage, wholesale, and processing levels, usually attributed to structural inadequacies across different countries, was estimated to be 13.2%, a percentage unchanged from 2016 and far from the target of halving postharvest food losses by 2030 (https://sdgs.un.org/goals/goal12#progress_and_info, 2024). In Europe, Directive 2008/98/EC of the European Parliament last modified in February 2024 defined the term "food waste" as "all food product that has become waste." Moreover, one object or substance can be considered a by-product if the following conditions are met: "(1) further use of the substance or object is certain, (2) the substance or object can be used directly without any further processing other than normal industrial practice, (3) the substance or object is produced as an integral part of a production

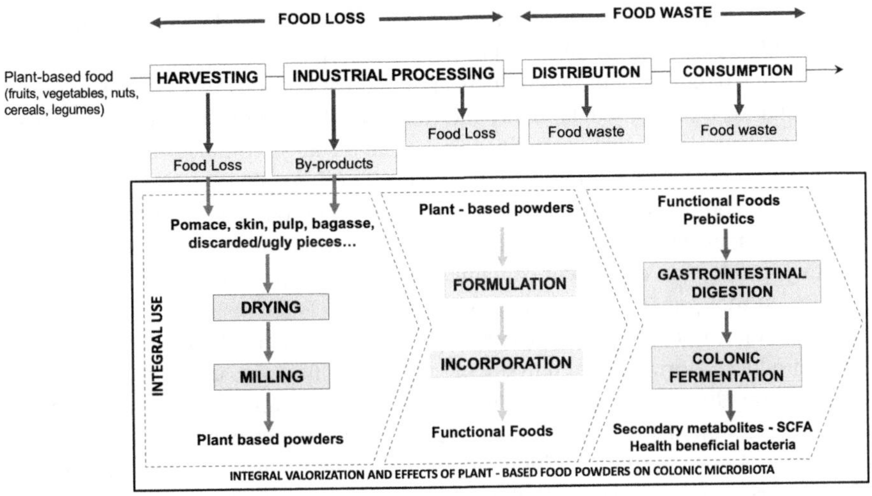

Fig. 3.1 Concepts of "food loss" and "food waste" along the food supply chain. Integral valorization of food losses and by-products through drying and milling into plant-based powders with prebiotic effects on colonic microbiota. *SCFA* short-chain fatty acid <original to the authors>

process, and (4) further use is lawful, i.e., the substance or object fulfills all relevant product, environmental, and health protection requirements for the specific use and will not lead to overall adverse environmental or human health impacts" (Fig. 3.1).

The EU platform on food losses and food waste (2024) has been created to support all actors in "defining measures needed to prevent food loss and food waste, sharing the best practices, and evaluating progress made over time." Quantification of European food loss and food waste levels (latest estimated by Eurostat 2022) revealed that 69% of EU food waste arises in the household, food service, and retail sectors, with production and processing sectors contributing to the remaining 31% of food loss (https://food.ec.europa.eu/safety/food-waste/eu-actions-against-food-waste/eu-platform-food-losses-and-food-waste_en). According to Santos et al. (2022), fruits and vegetables are responsible for about 22% of food losses along the supply chain. To date, no official data have been found on food losses associated with nuts and cereals. By-products generated by fruit and vegetable processing include mainly peels, cores, pomaces, and unripe, damaged, or "ugly" products that do not meet market standards (Padayachee et al. 2017). Fruit and vegetable industrial by-products/losses are typically secondary products from primary food production processes (Faustino et al. 2019) and are generally obtained from the industries of fresh and cut produce, canning, juice/nectar processing, wine production, or olive oil extraction (Santhosh et al. 2021). In the case of cereals and nuts, food losses are mainly the result of refining or removal of fibrous parts, extraction of oils, or production of plant-based beverages. Beverages from cereals, nuts, and oilseeds are increasing in popularity. Raw materials include soybean, rice, quinoa, almonds, walnuts, coconut, or tiger nuts. In the production of drinks, the solid material that is left is usually called press cake or bagasse and can represent from 10% to 50% of the raw product weight (Lorente et al. 2023).

Fruit, vegetable, and nut by-products are good natural sources of carbohydrates, polysaccharides, proteins, fats, fiber, and biomolecules such as vitamins, minerals, and antioxidants. Their high humidity content and the large volumes obtained during processing are a problem for manufacturers. The low pH and the presence of antibacterial compounds such as phenols can limit their degradation (Antonić et al. 2020). The main limitation, however, is that they need to be treated immediately after their production to prevent contamination or microbial growth.

Some current practices for dealing with lost products are not only costly but also harmful to the environment. There is a clear interest in the development of strategies for by-product valorization, reintroducing them into the value chain in an environmentally friendly way. For example, sustainable economic development can be achieved through the valorization or creation of inexpensive by-products. It is essential to know the composition of the considered by-products to define the possible uses and to establish the most important valorization opportunities. By-product composition will determine their nutritional and technological properties. Nutritional and dietary value, water and oil interaction properties, foaming capacity, and other properties relevant to food industry applications are determined by the type and amount of fiber, the lipid profile, and the type and percentage of protein (Lorente et al. 2023). Similarly, nutraceutical applications will be determined by the type and amount of bioactive compounds, such as antioxidants. In any case, the inclusion of natural ingredients, coming from fruit, vegetable, or nut processing, in staple foods could be more acceptable to most people than their synthetic counterparts (Kowalska et al. 2017).

3.1.2 Valorization Strategies, By-Product Composition, and Uses

Some of the processing operations being investigated to increase the potential for valorization of plant-based by-products include solid-state fermentation, extraction processes, and dehydration techniques. These are operations that can be implemented in a sustainable way, using renewable energy and without the use of solvents or other polluting substances.

Solid-state fermentation using fungi, yeasts, or bacteria enables the production of value-added compounds such as enzymes, amino acids, biopesticides, biofuels, biosurfactants, organic acids, flavors, colorings, aromatic compounds, biologically active secondary metabolites, and other substances of interest to the food industry. It applies mainly to residues rich in carbohydrates, fiber, and a small amount of protein. The various fruit pomace/press cakes are a good source of fiber, especially the insoluble fraction (Amaya-Cruz et al. 2015; Rocha Parra et al. 2015). Thus, apple pomace is a rich source of pectin, carbohydrates, crude fiber, and minerals, and it was fermented by Rodriguez Madera et al. (2016) using three indigenous cider yeast strains (*Saccharomyces cerevisiae*, *Hanseniaspora valbyensis*, and *Hanseniaspora uvarum*) and a commercial *S. cerevisiae* in combination with an enzymatic preparation as inoculum. In quantitative terms, the production of flavors was strongly dependent on the strain. Higher levels of fatty acids and their corresponding ethyl esters were accumulated in apple pomace inoculated with *Saccharomyces* strains. In the same way, other fruit by-products such as dried mango by-products can be used as a substrate for pectinase production by microorganisms such as *Fusarium* spp. (Reddy and Saritha 2015) as they contain significant amounts of pectin, other carbohydrates, and proteins and have low fat content.

Extraction processes are applied with the aim of obtaining antimicrobial extracts, proteins, or other specific bioactive compounds that can serve as additives for the food industry. This requires the presence of these compounds in the by-product of interest and the use of environmentally friendly solvents or sustainable treatments. Thus, fruit and vegetable by-products may serve as potential natural antimicrobials, mainly due to their richness in antioxidant and radical scavenging compounds. Apple by-products are also a source of phenolic compounds such as chlorogenic acid, phloridzin, epicatechin, catechin, and anthocyanins (Rabetafika et al. 2014). By-products from olive oil production have high antioxidant activity and a characteristic profile of fatty acids, amino acids, and mineral ions, such as sodium, potassium, calcium, and magnesium (Rodrigues et al. 2015). Peels, filter cakes, and sediment from pomegranate nectar production contained more phenolic compounds and exhibited higher antioxidant activity than those obtained from pomegranate nectar (Surek and Nilufer-Erdil 2016). Avocado fruit seeds are rich in not only fiber but also antioxidants, antimicrobials, colorants, flavorings, and thickening agents (Barbosa-Martín et al. 2016). Their seeds contain various components such as saponin, phytosterols, triterpenes, fatty acids, furoic acids, flavonol dimers, and proanthocyanidins. Some of them are related to antimicrobial and antifungal activities and larvicidal effects (Barbosa-Martín et al. 2016; Rodríguez-Carpena et al. 2011). Regarding protein extraction, protein isolate obtained from tomato seed contains all the essential amino acids (including lysine) and meets the minimum requirements of the reference protein for children aged from 1 to 2 years (Sarkar and Kaul 2014).

Drying by-products by hot air or sun is one of the most technologically simple, economical, and sustainable valorization strategies. The process does not produce waste or solvents, and it involves the integral use of the by-product, which is converted into flour or powder with the possibility of varying particle size by milling depending on the intended use. The resulting product is a plant-based powder or flour rich in nutrients and/or antioxidant compounds that can be used as an ingredient in other food formulations (Fig. 3.1). As an example, flour obtained from guava by-products can be useful in the production of cakes with high concentrations of biocomponents. A cake recipe with different proportions of flour obtained from guava peel (30, 50, and 70%) was developed by Bertagnolli et al. (2014). The flour addition was characterized by a high content of dietary fiber, minerals, polyphenols, and β-carotene. In addition, the cakes obtained were characterized by a low content of lipids and carbohydrates. In addition, baru (Brazilian almond) flour from oil extraction was used as a substitute for wheat flour in the production of biscuits, with a positive effect on the nutritional and antioxidant properties of the biscuits

(Pineli et al. 2015). On the other hand, nut flour from bagasse is rich in fat and protein and could be used to fortify plant protein formulations.

However, the management of process variables such as temperature and drying time, as well as the initial and final characteristics of the by-products, can influence the composition and the functional and technological properties of the flours obtained. The effects of drying on the functional and technological properties of fruit and vegetable by-products have been recently reviewed by Santos et al. (2022) and Ritika et al. (2024).

Generally, drying temperatures used in by-products and reported in the literature were mild (50–60 °C) to minimally affect antioxidant compounds. Mild drying temperatures and lower humidity products resulted in a higher content of bioactive compounds and attractive colors in soybean okara, wheat germ, banana, and spent coffee grounds (Yang et al. 2020). Lower temperatures (30–40 °C) implied higher dehydration times and more oxidation reactions. On the contrary, higher temperatures (70–90 °C) implied shorter dehydration times but more temperature effects such as protein denaturation and cell membrane disintegration, higher Maillard reaction products, and higher antioxidant compound losses. Additionally, higher air-drying temperatures (70–90 °C) resulted in lower water retention capacity and fat adsorption capacity in orange by-products (Garau et al. 2007). Regarding antioxidant activity, orange by-products air-dried at 60 °C had higher values (Garau et al. 2007). Fruit and vegetable by-product flours are especially rich in fibers that can be bound to bioactive compounds such as phenolic acids and carotenoids (Acosta-Estrada et al. 2014). This fiber–bioactive compound bound can be formed due to product structure changes that occur during drying. Zlatanović et al. (2019) noted a higher level of phenolics bound to larger particles of apple pomace flours. When bound, bioactive compounds can be effectively delivered in the gut as the fiber acts as a carrier for their transport along the gastrointestinal tract. After fiber fermentation by the gut microbiota, bioactive compounds can be then released (Acosta-Estrada et al. 2014; Gómez and Martínez 2018). In nuts, denaturation of proteins can result in a similar effect, protecting antioxidant compounds from degradation during processing, but also affecting their bioaccessibility and releasing them in the gut. Duarte et al. (2024) observed that throughout the dehydration process, the antioxidant compounds in almond bagasse were preserved without significant differences among dehydration treatments. However, due to the fat content, rancidity of fats during storage could be the main problem of almond bagasse flour. Duarte et al. (2024) found that the acidity index and peroxide index were increased in 180 days of storage period, although they did not reach critical values. Dehydrated samples stored at room temperature showed peroxide values ranging from 10 to 12.8 meq O2/kg dry matter. Meanwhile, those samples subjected to accelerated storage showed peroxide values from 14.7 to 23 meq O2/kg dry matter.

3.2 Gut Microbiota and Diet

3.2.1 Interactions Between Foods and the Gut Microbiota

The human intestinal microbiome is a complex and dynamic ecosystem that has coevolved with humans and has established close symbiotic relationships, regulating multiple aspects of human physiology (Gilbert et al. 2018; Van de Guchte et al. 2018). The gastrointestinal tract provides stable growth conditions, in terms of temperature and pH, and a constant source of nutrients for the microbial community. For its part, the intestinal microbiota carries out a wide variety of functions that benefit the host, such as the synthesis of essential vitamins and amino acids, and participates in the development and regulation of the immune system; the proliferation, differentiation, and maintenance of the intestinal epithelium; protection against pathogens; or regulation of the nervous system (Flint et al. 2012; Huda et al. 2022). Likewise, the microbiota plays a fundamental role in nutrition, intervening, among others, in different metabolic processes such as the degradation of complex polysaccharides and fiber or in the metabolism of polyphenols and carotenoids since these compounds reach the colon partially degraded or undigested (Flint 2012; Mosele et al. 2015). Thus, they are metabolized by the intestinal microbiota, generating structurally simpler metabolites, such as short-chain fatty acids (SCFA), that are easily absorbed and exported to other organs. The SCFAs—acetate, butyrate, and propionate—are well recognized to interact with different host systems (epithelial, immune, nervous, and endocrine) and facilitate many beneficial health effects. The profile of the microbial metabolites produced with beneficial health effects will depend on the composition of the intestinal microbiota. In turn, the structure and function of the gut microbiota are influenced by various factors with greater or lesser selective force such as age, genotype, lifestyle, health history, and diet (Gilbert et al. 2018). Hence, the microbiota composition presents a high interindividual variability, though some key bacterial species are typically present in most of the individuals (Turnbaugh et al. 2009; Human Microbiome Project Consortium 2012; Qin et al. 2010). Different works on the interactions between diet and microbiota pointed out that the diet explains over 20% and 50% of this microbial structural variation in humans and mice, respectively (Ussar et al. 2015; Rothschild et al. 2018; Leeming et al. 2019; Beam et al. 2021). In fact, the diet is the major driving force for modulating the gut microbiota. Sufficiently drastic short-term dietary interventions induce a shift in the gut microbiota, but the changes appear to be temporary, not persisting for more than a few days (David et al. 2014a, David et al. 2014b). However, long-term diet has very large effects on gut microbiome composition (Wu et al. 2011). Thus, it has been seen that different diets, such as gluten-free, Western, vegetarian, and Mediterranean, modulate the microbiota in different ways, having an impact on overall health. Broadly, the consumption of animal-based protein promotes the growth of bile-tolerant bacteria such as *Bacteroides*, *Alistipes*, and *Bilophila*, while plant-based protein intake increases the abundance of the gut commensals *Bifidobacterium* and *Lactobacillus* and the production of SCFAs. Similarly, animal and vegetable fats have a strong effect on the composition of the microbiota, and differences have even been found in the modulation of the microbiota between saturated and unsaturated fats (Lerma-Aguilera et al. 2024; Singh et al. 2017). Asnicar et al. (2021) highlighted a strong association between a butyrate-producing bacterium, *Lawsonibacter asaccharolyticus*, and coffee consumption and an increased abundance of *Bifidobacterium animalis* and *Streptococcus thermophilus* with full-fat yogurt consumption. Nondigestible carbohydrates are the most well-studied dietary compounds in relation to the modulation of the gut microbiota and the effects of their secondary metabolites, the SCFAs, on different physiological, metabolic, and immune processes in the human body.

Nutritional interventions have been carried out to correct the alteration of the microbiota associated with different pathologies such as obesity, type II diabetes, HIV, hypercholesterolemia, or inflammatory bowel disease (IBD) (Gebrayel et al. 2022). However, these dietary strategies have not always been as effective as desired, since in many cases the response is specific to each individual due to the great interindividual variability that exists in the composition of the microbiota. Likewise, the heterogeneity in responsiveness to the intake of plant bioactive compounds can hamper the identification of the beneficial effects of these compounds in human studies. In recent years, both in vitro and in vivo studies have addressed the associations between specific foods and the response of specific bacterial genera, considering the interindividual variability of gut microbiota composition. Therefore, it is necessary to move toward personalized nutrition in which the intestinal microbiome is taken into account to obtain the best response and the greatest health benefits.

3.2.2 Gut Simulators

Currently, there is extensive scientific evidence of the close relationship between diet, microbiota, and health (de Vos et al. 2022). However, elucidating the interaction mechanisms among them requires a precise understanding of the gastrointestinal digestion and colonic fermentation processes through which foods and bioactive compounds are transformed and interact with the human body to exert their eventual biological functions. Therefore, it is necessary to clarify which intestinal bacteria degrade a bioactive compound, which metabolites they produce, and which are the metabolic pathways used. This challenge has prompted the development of in vitro models that simulate gastrointestinal digestion and colonic fermentation of food. In vitro digestion studies address research questions such as the digestibility of macronutrients, the bioaccessibility of bioactive compounds, the release of encapsulated substances, or the effect of gastrointestinal conditions on digestibility, among others. In relation to the microbiota, these models, combined with next-generation sequencing technology, can be useful to evaluate the effects of different nutrients, foods, additives, prebiotics, probiotics, or synbiotics on the microbiota composition and functions.

In vitro intestinal fermentation models range from static batch fermentations, which are not only the simplest but also the most versatile, the most economical, and the easiest to apply at the laboratory level, to sophisticated pH-controlled multistage continuous culture systems. Among the latter, the Simulator of the Human Intestinal Microbial Ecosystem (SHIME®) and TNO in vitro model of the colon (TIM-2) models are the most widely applied, although the dynamic gastrointestinal simulator (SIMGI®), CoMiniGut, and PolyFermS simulators are also commonly used (Veintimilla-Gozalbo et al. 2021; Moon et al. 2016; Singh et al. 2022). Most of these models do not include the interactions between the host and the microbiota at the physiological or immune levels. The continuous systems better simulate the physiological gut conditions and also better represent the growing intestinal microbiota during long time periods than the static batch fermenters. However, they have limited statistical power when processing a low number of samples. On the other hand, in vitro gut models allow estimating the levels and the ratio of SCFA production from different substrates more accurately than their in vivo quantification from feces due to the rapid absorption of SCFAs by colonocytes.

3.2.3 Plant-Based Ingredients as Prebiotics

Prebiotics are currently defined as "a substrate that is selectively utilized by host microorganisms conferring a health benefit" (Gibson et al. 2017). Among all food ingredients, nondigestible carbohydrates (oligosaccharides and polysaccharides) are the most important candidates to be considered as prebiotics. The first studies on prebiotics date back to the 80 s and were performed on fructans (fructo-oligosaccharides [FOS]) and galactans (galacto-oligosaccharides [GOS]) derived from plants (Roberfroid et al. 1998). However, other nondigestible carbohydrates such as xylooligosaccharides, β-glucans, and macroalgal polysaccharides are currently studied, and dosage-dependent effects have been detected (Cunningham et al. 2021). Currently, agricultural waste, surpluses, and by-products are being considered new sources of prebiotics through their valorization, since they present an important nutritional richness. In many cases, they are rich, in addition to fiber, in micronutrients such as polyphenols and carotenoids or in macronutrients such as proteins and polyunsaturated fats.

Prebiotic effects may be mediated directly through the metabolism of the compounds by specific bacterial groups, promoting their growth and metabolic activity, or indirectly by cross-feeding, since the degradation of complex polysaccharides involves a complex trophic network of microorganisms. The key health-promoting microbial metabolites are the SCFAs, mainly butyrate, acetate, and propionate, and they are involved in a range of functions in the colon and other organs such as the liver and the brain, including gut epithelial and mucus barrier function, innate and adaptative immunity, inflammation, blood pressure, glucose and lipid metabolism, energy expenditure, and appetite regulation (Blaak et al. 2020). A recent investigation has shown that differences in dietary fiber properties, such as length, branching, solubility, or charge, affect their impact on the microbiota composition and functions. Thus, fiber from whole grain products increases acetic and propionic acids by promoting the growth of bifidobacteria and lactobacilli in short-term fermentations, while psyllium fiber had the highest effect by stimulating the growth of *Bacteroidota* together with the production of propionate (Calatayud et al. 2021). Likewise, differences in the chemical structure of pectin could have a significant influence on fermentation outcomes, microbiota structure, and SCFA profiles (Firrman et al. 2022; Lancaster et al. 2022). Recently, Mitsou et al. (2020) suggested the beneficial effects of mushroom β-glucans on gut microbiota and SCFA production using fecal inocula from elder subjects. Colonic fermentation studies with weaning infant feces as inocula showed that resistant starch promotes the growth of several commensal genera such as *Prevotella*, *Veillonella*, *Bifidobacterium*, and *Collinsella*, as well as propionate production (Kalhan 2018; Wang et al. 2019). These results suggested that the supplementation with this polysaccharide might have benefits for infant health during weaning. The understanding of fiber effects on gut microbiota could lead to potential health promotion through dietary fiber diversification.

On the other hand, different compounds, such as vitamins, minerals, fatty acids, carotenoids, or phenolic compounds, present prebiotic properties (Pham et al. 2021; Cunningham et al. 2021). For decades, carotenoids and polyphenols have been studied in relation to health due to their anti-inflammatory and antioxidant properties, focusing on the parts of the phytochemicals that, after being ingested, enter the circulatory system. However, due to their complex structures and high molecular weights, only a small portion of what is ingested is absorbed; the rest arrives in the colon where it is metabolized by the microbiota. The biological effects of microbial metabolites produced in the colon from both carotenoids and polyphenols have not been thoroughly investigated. Different studies have suggested prebiotic-like effects of carotenoids such as bacterial shifts with health-associated properties, stimulation of IgA, or stabilization of the mucosal tight junctions (Eroglu et al. 2023). Also, mice studies have shown that physiological amounts of dietary β-carotene reduce adipose tissue size, promote adipose tissue browning, and favor an overall induction of fatty acid oxidation in adipocytes (Coronel et al. 2019). In a clinical trial, lycopene supplementation resulted in an increase in the abundance of different colonic *Bifidobacterium* and *Lactobacilli* species as well as an improvement in gut metabolism (Wiese et al. 2019). Recently, Dai et al. (2022) showed in a fermentation model study that the four typical carotenoids, β-carotene, lutein, lycopene, and astaxanthin, have a differential impact on the microbiota structure and all of them significantly increase the total SCFA production. Likewise, polyphenols have been extensively studied in relation to their strong antioxidant and anti-inflammatory properties and their modulating effect on the gut microbiota (Carrera-Quintanar et al. 2018). Polyphenols and carotenoids present in powder ingredients obtained from persimmon by-products and blueberry bagasse affect the microbial community growing during in vitro colonic fermentation, resulting in specific microbial communities (Bas-Bellver et al. 2020). Recently, in vivo and in vitro studies have shown that dietary supplementation with soy isoflavone modulates the microbiota in menopausal women, increasing bacteria involved in

SCFA production (Guadamuro et al. 2017, 2021). In an in vitro digestion/fermentation study (Duarte et al. 2023), almond bagasse stabilization via a dehydration treatment combined with milling was shown to be an alternative for producing potential prebiotics with high content of fiber and strong antioxidant properties (Fig. 3.1). Moreover, the potential almond prebiotics impacted the bacteria growing in the colonic fermentation, promoting SCFA-producing bacteria. Tannins are a large class of polyphenols with a wide range of chemical compositions and molecular weights that are produced by different families of higher plants. These compounds present, among others, marked antioxidant and antimicrobial properties, so they are used as preservative agents in foods (Molino et al. 2023). Recently, Molino et al. (2021) observed in in vitro fermentation studies that tannin extracts added to various foods favored the growth of the genus *Akkermansia*, considered as a marker of gut homeostasis, and of SCFA-producing members of the *Lachnospiraceae* and *Ruminococcaceae* families.

In the next years, the field of prebiotics will be expanded to include both carbohydrates and noncarbohydrate substances of plant, animal, or fungal origin. Furthermore, their application and health benefits will not be restricted to the digestive system but to other parts of the human body such as the skin, oral cavity, or vagina.

Acknowledgments This study was supported by MCIN with funding from the European Union NextGenerationEU (PRTR-C17.I1) and by Generalitat Valenciana (AGROALNEXT-2022/039 and CIAICO/2022/027).

References

Acosta-Estrada BA, Gutiérrez-Uribe JA, Serna-Saldívar SO (2014) Bound phenolics in foods, a review. Food Chem 152:46–55. https://doi.org/10.1016/j.foodchem.2013.11.093

Amaya-Cruz DM, Rodríguez-González S, Pérez-Ramírez IF et al (2015) Juice by-products as a source of dietary fibre and antioxidants and their effect on hepatic steatosis. J Funct Foods 17: 93–102. https://doi.org/10.1016/j.jff.2015.04.051

Antonić B, Jančíková S, Dordević D et al (2020) Grape pomace valorization: a systematic review and meta-analysis. Food Secur 9:1627. https://doi.org/10.3390/FOODS9111627

Asnicar F, Berry SE, Valdes AM et al (2021) Microbiome connections with host metabolism and habitual diet from 1,098 deeply phenotyped individuals. Nat Med 27:321–332. https://doi.org/10.1038/s41591-020-01183-8

Barbosa-Martín E, Chel-Guerrero L, González-Mondragón E et al (2016) Chemical and technological properties of avocado (Persea americana Mill.) seed fibrous residues. Food Bioprod Process 100(A):457–463. https://doi.org/10.1016/j.fbp.2016.09.006

Bas-Bellver C, Andrés C, Seguí L (2020) Valorization of persimmon and blueberry by-products to obtain functional powders: in vitro digestion and fermentation by gut microbiota. J Agric Food Chem 29:8080–8090. https://doi.org/10.1021/acs.jafc.0c02088

Beam A, Clinger E, Hao L (2021) Effect of diet and dietary components on the composition of the gut microbiota. Nutrients 15:2795. https://doi.org/10.3390/nu13082795

Bertagnolli SMM, Silveira MLR, Fogaca AD et al (2014) Bioactive compounds and acceptance of cookies made with Guava peel flour. Ciênc Tecnol Aliment 34:303–308. http://www.scielo.br/pdf/cta/v34n2/13.pdf

Blaak EE, Canfora EE, Theis S et al (2020) Short chain fatty acids in human gut and metabolic health. Benef Microbes 1:411–455. https://doi.org/10.3920/BM2020.0057

Calatayud M, Van den Abbeele P, Ghyselinck J et al (2021) Comparative effect of 22 dietary sources of fiber on gut microbiota of healthy humans in vitro. Front Nutr 2:700571. https://doi.org/10.3389/fnut.2021.700571

Carrera-Quintanar L, López Roa RI, Quintero-Fabián S et al (2018) Phytochemicals that influence gut microbiota as prophylactics and for the treatment of obesity and inflammatory diseases. Mediat Inflamm 26:9734845. https://doi.org/10.1155/2018/9734845

Coronel J, Pinos I, Amengual J et al (2019) β-carotene in obesity research: technical considerations and current status of the field. Nutrients 11:842. https://doi.org/10.3390/nu11040842

Cunningham M, Azcarate-Peril MA, Barnard A et al (2021) Shaping the future of probiotics and prebiotics. Trends Microbiol 29:667–685. https://doi.org/10.1016/j.tim.2021.01.003

Dai Z, Li Z, Shi E et al (2022) Study on the interaction between four typical carotenoids and human gut microflora using an in vitro fermentation model. J Agric Food Chem 26:13592–13601. https://doi.org/10.1021/acs.jafc.2c03464

David LA, Materna AC, Friedman J (2014a) Host lifestyle affects human microbiota on daily timescales. Genome Biol 15:R89. https://doi.org/10.1186/gb-2014-15-7-r89

David LA, Maurice CF, Carmody RN et al (2014b) Diet rapidly and reproducibly alters the human gut microbiome. Nature 23:559–563. https://doi.org/10.1038/nature12820

de Vos WM, Tilg H, van Hul M et al (2022) Gut microbiome and health: mechanistic insights. Gut 71:1020–1032. https://doi.org/10.1136/gutjnl-2021-326789

Directive 2008/98/EC of the European parliament and of the council of 19 November 2008 on waste and repealing certain Directives (Text with EEA relevance). Current consolidated version: 18/02/2024. ELI: http://data.europa.eu/eli/dir/2008/98/oj. Accessed on 04/06/2024

Duarte S, Puchades A, Jiménez-Hernández N et al (2023) Almond (*Prunus dulcis*) bagasse as a source of bioactive compounds with antioxidant properties: an in vitro assessment. Antioxidants (Basel) 7:1229. https://doi.org/10.3390/antiox12061229

Duarte S, Betoret E, Betoret N (2024) Shelf life and functional quality of almond bagasse powders as influenced by dehydration and storing conditions. Food Secur 13:744. https://doi.org/10.3390/foods13050744

EPA (2022) Sustainable management of food: food recovery hierarchy. United States Environmental Protection Agency. https://www.epa.gov/sustainable-managementfood/food-recovery-hierarchy

Eroglu A, Al'Abri IS, Kopec RE et al (2023) Carotenoids and their health benefits as derived via their interactions with gut microbiota. Adv Nutr 14:238–255. https://doi.org/10.1016/j.advnut.2022.10.007

EU platform on food losses and food waste.: https://food.ec.europa.eu/safety/food-waste/eu-actions-against-food-waste/eu-platform-food-losses-and-food-waste_en. Accessed on 04/06/2024

Eurostat 2022. https://EC.europa.eu/eurostat/statisticsexplained/index.php?title=Food_waste_and_food_waste_prevention_-_estimates&stable=0&redirect=no. Accessed on 04/06/2024

FAO (2019) The state of food and agriculture 2019. Moving forward on food loss and waste reduction. Rome. Licence: CC BY-NC-SA 3.0 IGO

Faustino M, Veiga M, Sousa P et al (2019) Agrofood by-products as a new source of natural food additives. Molecules 24:1056. https://doi.org/10.3390/MOLECULES24061056

Firrman J, Mahalak K, Bobokalonov J et al (2022) Modulation of the gut microbiota structure and function by two structurally different lemon pectins. Food Secur 1:3877. https://doi.org/10.3390/foods11233877

Flint HJ (2012) The impact of nutrition on the human microbiome. Nutr Rev 70(Suppl 1):S10. https://doi.org/10.1111/j.1753-4887.2012.00499.x

Flint HJ, Scott KP, Duncan SH et al (2012) Microbial degradation of complex carbohydrates in the gut. Gut Microbes 3:289–306. https://doi.org/10.4161/gmic.19897

Garau MC, Simal S, Rossello C et al (2007) Effect of air-drying temperature on physico-chemical properties of dietary fibre and antioxidant capacity of orange (Citrus aurantium v. Canoneta) by-products. Food Chem 104:1014–1024. https://doi.org/10.1016/j.foodchem.2007.01.009

Gebrayel P, Nicco C, Al Khodor S et al (2022) Microbiota medicine: towards clinical revolution. J Transl Med 7:111. https://doi.org/10.1186/s12967-022-03296-9

Gibson GR, Hutkins R, Sanders ME et al (2017) Expert consensus document: the International Scientific Association for Probiotics and Prebiotics (ISAPP) consensus statement on the definition and scope of prebiotics. Nat Rev Gastroenterol Hepatol 14:491–502. https://doi.org/10.1038/nrgastro.2017.75

Gilbert JA, Blaser MJ, Caporaso JG et al (2018) Current understanding of the human microbiome. Nat Med 24:392–400. https://doi.org/10.1038/nm.4517

Gómez M, Martínez MM (2018) Fruit and vegetable by-products as novel ingredients to improve the nutritional quality of baked goods. Crit Rev Food Sci Nutr 58:2119–2135. https://doi.org/10.1080/10408398.2017.1305946

Guadamuro L, Dohrmann AB, Tebbe CC et al (2017) Bacterial communities and metabolic activity of faecal cultures from equol producer and non-producer menopausal women under treatment with soy isoflavones. BMC Microbiol 17:93. https://doi.org/10.1186/s12866-017-1001-y

Guadamuro L, Azcárate-Peril MA, Tojo R et al (2021) Impact of dietary isoflavone supplementation on the fecal microbiota and its metabolites in postmenopausal women. Int J Environ Res Public Health 18:7939. https://doi.org/10.3390/ijerph18157939

Huda MN, Salvador AC, Barrington WT et al (2022) Gut microbiota and host genetics modulate the effect of diverse diet patterns on metabolic health. Front Nutr 18:896348. https://doi.org/10.3389/fnut.2022.896348

Human Microbiome Project Consortium (2012) Structure, function and diversity of the healthy human microbiome. Nature 13:207–214. https://doi.org/10.1038/nature11234

Kalhan SC (2018) Microbial fermentation of starch: its impact on the range of acceptable carbohydrate intake. J Pediatr Gastroenterol Nutr 66(Suppl 3):S42–S45. https://doi.org/10.1097/mpg.0000000000001827

Kowalska H, Czajkowska K, Cichowska J et al (2017) What's new in biopotential of fruit and vegetable by-products applied in the food processing industry. Trends Food Sci Technol 67:150–159. https://doi.org/10.1016/j.tifs.2017.06.016

Lancaster SM, Lee-McMullen B, Abbott CW et al (2022) Global, distinctive, and personal changes in molecular and microbial profiles by specific fibers in humans. Cell Host Microbe 30:848–862. https://doi.org/10.1016/j.chom.2022.03.036

Leeming ER, Johnson AJ, Spector TD et al (2019) Effect of diet on the gut microbiota: rethinking intervention duration. Nutrients 22:2862. https://doi.org/10.3390/nu11122862

Lerma-Aguilera AM, Pérez-Burillo S, Navajas-Porras B et al (2024) Effects of different foods and cooking methods on the gut microbiota: an in vitro approach. Front Microbiol 8:1334623. https://doi.org/10.3389/fmicb.2023.1334623

Lorente D, Duarte S, Betoret E et al (2023) Opportunities for the valorization of waste generated by the plant-based milk substitutes industry. In: Advanced technologies in waste water treatment: food processing industry, pp 25–66

Mitsou EK, Saxami G, Stamoulou E et al (2020) Effects of rich in B-glucans edible mushrooms on aging gut microbiota characteristics: an in vitro study. Molecules 25:2806. https://doi.org/10.3390/molecules25122806

Molino S, Lerma-Aguilera A, Jiménez-Hernández N et al (2021) Enrichment of food with tannin extracts promotes healthy changes in the human gut microbiota. Front Microbiol 16:625782. https://doi.org/10.3389/fmicb.2021.625782

Molino S, Francino MP, Rufián JA (2023) Why is it important to understand the nature and chemistry of tannins to exploit their potential as nutraceuticals? Food Res Int 173:113329. https://doi.org/10.1016/j.foodres.2023.113329

Moon JS, Li L, Bang J, Han NS (2016) Application of in vitro gut fermentation models to food components: a review. Food Sci Biotechnol 25(Suppl 1):1–7. https://doi.org/10.1007/s10068-016-0091-x

Mosele JI, Macià A, Motilva MJ (2015) Metabolic and microbial modulation of the large intestine ecosystem by non-absorbed diet phenolic compounds: a review. Molecules 18:17429–17468. https://doi.org/10.3390/molecules200917429

Padayachee A, Day L, Howell K et al (2017) Complexity and health functionality of plant cell wall fibers from fruits and vegetables. Crit Rev Food Sci Nutr 57:59–81. https://doi.org/10.1080/10408398.2013.850652

Pham VT, Fehlbaum S, Seifert N et al (2021) Effects of colon-targeted vitamins on the composition and metabolic activity of the human gut microbiome- a pilot study. Gut Microbes 13:1–20. https://doi.org/10.1080/19490976.2021.1875774

Pineli L d L d O, Veras de Carvalho M, Andrade de Aguiar L et al (2015) Use of baru (Brazilian almond) waste from physical extraction of oil to produce flour and cookies. LWT- Food Sci Technol 60:50–55. https://doi.org/10.1016/j.lwt.2014.09.035

Qin N, Yang H, Wang J et al (2010) A human gut microbial gene catalogue established by metagenomic sequencing. Nature 4:59–65. https://doi.org/10.1038/nature08821

Rabetafika HN, Bchir B, Blecker C et al (2014) Fractionation of apple by-products as source of new ingredients: current situation and perspectives. Trends Food Sci Technol 40:99–114. https://doi.org/10.1016/j.tifs.2014.08.004

Reddy MP, Saritha KV (2015) Bio-catalysis of mango industrial waste by newly isolated *Fusarium* sp. (PSTF1) for pectinase production. Biotech 5:893–900. https://doi.org/10.1007/s13205-015-0288-3

Ritika R, Shukla S et al (2024) Valorisation of fruit waste for harnessing the bioactive compounds and its therapeutic application. Trends Food Sci Technol 144:104302. https://doi.org/10.1016/j.tifs.2023.104302

Roberfroid MB, Van Loo JA, Gibson GR (1998) The bifidogenic nature of chicory inulin and its hydrolysis products. J Nutr 128:11–19. https://doi.org/10.1093/jn/128.1.11

Rocha Parra AF, Ribotta PD, Ferrero C (2015) Apple pomace in gluten-free formulations: effect on rheology and product quality. Int J Food Sci Tech 50:682–690. https://doi.org/10.1111/ijfs.12662

Rodrigues F, Pimentel FB, Oliveira MBPP (2015) Olive by-products: challenge application in cosmetic industry. Ind Crop Prod 70:116–124. https://doi.org/10.1016/j.indcrop.2015.03.027

Rodriguez Madera R, Pando Bedrinana R, Suárez VB (2016) Production and characterization of aroma compounds from apple pomace by solid-state fermentation with selected yeasts. LWT-Food Sci Technol 64:1342–1353. https://doi.org/10.1016/j.lwt.2015.07.056

Rodríguez-Carpena JG, Morcuende D, Andrade MJ et al (2011) Avocado (*Persea americana Mill.*) phenolics, in vitro antioxidant and antimicrobial activities, and inhibition of li-pid and protein oxidation in porcine patties. J Agric Food Chem 59:5625–5635. https://doi.org/10.1021/jf1048832

Rothschild D, Weissbrod O, Barkan E et al (2018) Environment dominates over host genetics in shaping human gut microbiota. Nature 8:210–215. https://doi.org/10.1038/nature25973

Santhosh R, Nath D, Sarkar P (2021) Novel food packaging materials including plant-based byproducts: a review. Trends Food Sci Technol 118:471–489. https://doi.org/10.1016/J.TIFS.2021.10.013

Santos D, Lopes da Silva JA, Pintado M (2022) Fruit and vegetable by-products flours as ingredients: a review on production process, health benefits and technological functionalities. LWT-Food Sci Technol 154:112707. https://doi.org/10.1016/j.lwt.2021.112707

Sarkar A, Kaul P (2014) Evaluation of tomato processing by-products: a comparative study in a pilot scale setup. J Food Process Eng 37:299–307. https://doi.org/10.1111/jfpe.12086

Singh RK, Chang HW, Yan D et al (2017) Influence of diet on the gut microbiome and implications for human health. J Transl Med 8:73. https://doi.org/10.1186/s12967-017-1175-y

Singh V, Son H, Lee G et al (2022) Role, relevance, and possibilities of in vitro fermentation models in human dietary, and gut-microbial studies. Biotechnol Bioeng 119(11):3044–3061. https://doi.org/10.1002/bit.28206

Surek E, Nilufer-Erdil D (2016) Phenolic contents, antioxidant activities and potential bioaccessibilities of industrial pomegranate nectar processing wastes. Int J Food Sci Technol 51:231–239. https://doi.org/10.1111/ijfs.13000

Turnbaugh PJ, Hamady M, Yatsunenko T et al (2009) A core gut microbiome in obese and lean twins. Nature 22:480–484. https://doi.org/10.1038/nature07540

Sustainable development goals. https://sdgs.un.org/goals. Accessed on 03/06/2024

Sustainable development goals progress and info. https://sdgs.un.org/goals/goal12#progress_and_info. Accessed on 03/06/2024

Ussar S, Griffin NW, Bezy O et al (2015) Interactions between gut microbiota, host genetics and diet modulate the predisposition to obesity and metabolic syndrome. Cell Metab 1:516–530. https://doi.org/10.1016/j.cmet.2015.07.007

Van de Guchte M, Blottière HM, Doré J (2018) Humans as holobionts: implications for prevention and therapy. Microbiome 6:81. https://doi.org/10.1186/s40168-018-0466-8

Veintimilla-Gozalbo E, Asensio-Grau A, Calvo-Lerma J et al (2021) In vitro simulation of human colonic fermentation: a practical approach towards models' design and analytical tools. Appl Sci 11:8135. https://doi.org/10.3390/app11178135

Wang Y, Mortimer EK, Katundu KGH et al (2019) The capacity of the fecal microbiota from Malawian infants to ferment resistant starch. Front Microbiol 10:1459. https://doi.org/10.3389/fmicb.2019.01459

Wiese M, Bashmakov Y, Chalyk N et al (2019) Prebiotic effect of lycopene and dark chocolate on gut microbiome with systemic changes in liver metabolism, skeletal muscles and skin in moderately obese persons. Biomed Res Int 2:4625279. https://doi.org/10.1155/2019/4625279

Wu GD, Chen J, Hoffmann C et al (2011) Linking long-term dietary patterns with gut microbial enterotypes. Science 7:105–108. https://doi.org/10.1126/science.1208344

Yang H, Sombatngamwilai T, Yu WY et al (2020) Drying applications during value-added sustainable processing for selected mass-produced food coproducts. PRO 8:307. https://doi.org/10.3390/pr8030307

Zlatanović S, Kalušević A, Micić D et al (2019) Functionality and storability of cookies fortified at the industrial scale with up to 75% of apple pomace flour produced by dehydration. Food Secur 8:516. https://doi.org/10.3390/foods8110561

European Nutritional Guidelines for the Management of Childhood Obesity: The Role of Personalised Nutrition

4

Annarita Formisano, Giuseppe Iacomino, Marika Dello Russo, Paola Russo, Alfonso Siani, and Fabio Lauria

Abstract

Nutritional guidelines for managing obesity in children traditionally focus on limiting calorie intake, increasing physical activity, and promoting consumption of whole, unprocessed foods. However, recent scientific advances emphasise the critical role of the gut microbiota in regulating weight and metabolism, suggesting that nutritional interventions should also consider the diet's impact on gut microbial composition and function. The gut microbiota, an intricate ecosystem of microorganisms and their interactions, is one of science's most complex microbial communities.

While researchers have linked dietary modifications to various metabolic disorders, our understanding of the gut microbiome's role in metabolic health has primarily focused on quantifying changes in specific microbial groups, overlooking intricate interactions between species. To date, only a limited number of studies have explored precision nutrition approaches that integrate genetic profiles, microbiome composition, and physiological parameters to tailor dietary plans. Current nutritional guidelines for childhood obesity overlook critical aspects related to the gut microbiome, such as the influence of specific dietary components on microbial diversity, the potential of probiotics and prebiotics to modulate the gut microbiota, and the need for personalised microbiome analysis to account for individual variations in composition. Incorporating gut microbiota considerations into nutritional guidelines for childhood obesity promises a more comprehensive and effective approach to weight management, promoting long-term health and well-being.

A. Formisano · G. Iacomino · M. Dello Russo · P. Russo · A. Siani · F. Lauria (✉)
Institute of Food Sciences, National Research Council, Avellino, Italy
e-mail: flauria@isa.cnr.it

M. P. Francino, J. Á. Rufián Henares (eds.), *The Microbiome and Personalized Nutrition*, The Microbiomes of Humans, Animals, Plants, and the Environment 6,
https://doi.org/10.1007/978-3-031-93902-0_4

Keywords

European guidelines · Personalised nutrition · Obesity management · Childhood · Gut microbiome

4.1 Introduction

Obesity in children has been identified as a serious public health challenge that, over the past decades, has reached epidemic proportions in the European regions of the World Health Organization (WHO) (World Health Organization. Regional Office for Europe). Recent country-level estimates have indicated that one in eight school-aged children is affected by obesity, with large variations between countries (World Health Organization. Regional Office for Europe). This prevalence is lower in children aged 10–19 years, where one in four lives with overweight or obesity (Global Health Observatory n.d.; Bentham et al. 2017). For both age groups, the prevalence of obesity is higher in males and in the countries of the Mediterranean region (Global Health Observatory n.d.). The same data also indicate that the prevalence of obesity in children aged 5–19 years increased more rapidly, reaching approximately five times the 1975 levels by the year 2016. In the 10 years up to 2016, the prevalence of obesity increased by 40% (World Health Organization. Regional Office for Europe). Standardised data from the fifth round of the WHO European Region Childhood Obesity Surveillance Initiative (COSI) conducted between 2018 and 2020 and encompassing over 411,000 children aged 6–9 years across 33 countries have confirmed the high prevalence of obesity among children in this region and that it is more common in boys than girls. The data have further supported significant variation between countries, demonstrating a wide-ranging prevalence of obesity. The prevalence was highest in Cyprus, Italy, Greece, and Spain, and lowest in Denmark and Israel in both genders (World Health Organization. Childhood Obesity Surveillance Initiative 2021). In addition, in most countries, the prevalence of obesity has been higher in children whose parents had a low level of education than in those with high parental education (World Health Organization. Childhood Obesity Surveillance Initiative 2021). This association was particularly striking in several high-income countries, while in some middle-income countries, the direction of this relationship was reversed (World Health Organization. Childhood Obesity Surveillance Initiative 2021). Emerging evidence also suggests that shifts in food consumption and physical activity patterns determined by the COVID-19 pandemic may adversely affect childhood obesity trends (Pietrobelli et al. 2020; Androutsos et al. 2021). It is therefore feared that many WHO European countries will fail to meet the goal, identified by the World Health Assembly in 2012, of halting the rise in obesity by 2025 (Resolution 65.6) (Dadhich et al. 2019).

Obesity is a heterogeneous and multifactorial disease caused by genetic, environmental, metabolic, and psychosocial factors. This condition has been defined as an excess of adipose tissue that can induce a significant increase in health risks (World Health Organization 2022). In Europe, definitions of obesity by the WHO and by the

International Obesity Taskforce (IOTF) are the most widely used. In these definitions, the weight categories take into account the age- and gender-specific nutritional and developmental requirements of children. Therefore, in children, the variation of nutritional needs with age and development stage makes obesity treatment more complicated than in adults (World Health Organization 2022; Monasta et al. 2011). Childhood obesity is an important risk factor for short- and long-term health consequences (cardiorespiratory, cardiovascular, hepatic, and musculoskeletal) and for physical and psychosocial complications (Saha et al. 2011; Silva et al. 2018; Wühl 2019; O'Malley et al. 2021; Jebeile et al. 2022). Several studies have shown that obese children have lower school performance, cognitive ability, and self-esteem than normal-weight children, although many confounding factors such as sleep quality, nutritional status, and prejudice from teachers can influence this relationship (Khan et al. 2015, 2020; Martin et al. 2017; Gill et al. 2021). In addition, obesity can negatively affect a child's ability to play or engage in physical activities (Vandewater et al. 2015). In many cases, an obese child will be a future obese adult with a greater risk of developing chronic diseases related to excess weight, including cardiovascular disease, diabetes, and different types of cancer (Ayer et al. 2015; Weihe et al. 2020).

The increasing prevalence and long-term health consequences of childhood obesity have led many countries to design and implement programmes to prevent the early development of obesity in children. At the same time, there has been an increasing need to develop interventions for the treatment of children diagnosed with obesity. Diverse interventions for the treatment of paediatric obesity have been proposed (Mead et al. 2017; Rajjo et al. 2017; Singhal et al. 2021). Comprehensive approaches involving various elements—such as patient engagement, family counselling, promotion of physical activity, dietary adjustments, strategies for behaviour and lifestyle modification, and, when deemed appropriate, utilisation of drug therapies and surgical interventions—have gained extensive recognition as the fundamental approach to managing childhood obesity. Numerous systematic reviews have substantiated their effectiveness in the near term, leading to reductions in weight and body mass index (BMI) (Mead et al. 2017; Rajjo et al. 2017; Singhal et al. 2021). A meta-analysis showed that lifestyle interventions supervised by dieticians led to higher levels of relative weight loss than unsupervised ones (Duncanson et al. 2021). In the treatment of paediatric obesity, effective nutritional management is a crucial aspect. However, clinicians often struggle to stay updated with the extensive body of literature in this field. Clinical practice guidelines, along with position and consensus statements, provide valuable direction to enhance the care and management of children with obesity (Institute of Medicine U.S. 2011; Duncanson et al. 2021). These guidelines were formulated based on a combination of empirical evidence, previously published guidelines, clinical expertise, and expert opinions (Duncanson et al. 2021). Worldwide, numerous clinical guidelines have been developed and published by different paediatric societies or National Institutes for Health, over the past few decades (Delgado-Noguera et al. 2009; Institute of Medicine U.S. 2011; Duncanson et al. 2021). Several reviews have assessed the suitability and robustness of the guidance provided by these sources, revealing shared characteristics in recommendations related to the diagnosis and nutritional treatment of childhood obesity in primary care (Kirschenbaum and Gierut 2013; Richardson et al. 2013; Styne et al. 2017; Pfeifflé et al. 2019). However, these reviews have also indicated a lack of agreement regarding the methods for diagnosing and evaluating obesity in children (Delgado-Noguera et al. 2009; Kirschenbaum and Gierut 2013). Moreover, a range of studies have emphasised the significance of engaging the family to ensure the effectiveness of obesity management (Delgado-Noguera et al. 2009; Kirschenbaum and Gierut 2013). In a recent study, Pfeifflé et al. conducted an analysis and synthesis of recommendations commonly employed in clinical settings for managing the nutrition of obese children and adolescents. The authors reviewed 17 international guidelines and proposed a dietary strategy to facilitate weight loss (Pfeifflé et al. 2019). Their approach was based on fundamental principles of nutrition education, which included implementing a family-centred method and promoting the consumption of nutrient-rich foods while decreasing the intake of energy-dense alternatives, such as sugar-sweetened drinks. A primary dietary approach endorsed by several guidelines involved implementing a moderate reduction in energy intake (Michigan Quality Improvement Consortium (MQIC) 2018; Lee et al. 2016; Gow et al. 2017). Current findings propose that more intensive dietary strategies could yield distinct advantages, primarily for specific obesity-related complications. For instance, very low-energy diets (VLEDs) have been employed to effectively reverse the diagnosis of type 2 diabetes in young individuals (Willi et al. 2004; Gow et al. 2016). Similarly, both very low- and low-carbohydrate strategies have shown potential benefits for adolescents with insulin resistance and prediabetes (Goss et al. 2020; Vlassopoulos et al. 2021).

4.2 General Characteristics and Selection Criteria of the Main Guidelines for Nutritional Management and Dietary Intervention in Children with Obesity in Europe

The main clinical practice guidelines for the management of paediatric obesity are specific to each country in the European region (Kirschenbaum and Gierut 2013). Even if various national clinical guidelines have been developed, many European countries do not yet have their specific recommendations. Consequently, specialists frequently need to refer to guidelines from other countries or regions when devising obesity treatment strategies (Chen et al. 2022).

By consulting the websites of institutions involved in formulating clinical guidelines, such as the National Institute for Health and Clinical Excellence (NICE), and organisations dedicated to obesity, such as The Obesity Society and the European Association for the Study of Obesity (EASO), and utilising PubMed as a scientific research tool, a total of four clinical practice guidelines and two position and consensus statements specific to Europe were freely accessible in English. These resources, which met specific criteria (Spanish Ministry for Health 2009; Scottish Intercollegiate Guidelines Network 2010; National Institute for Clinical Excellence

2014, Johansen et al. 2015; Valerio et al. 2018; Hassapidou et al. 2023), were intended to be European guidelines and to include recommendations related to nutritional management and dietary interventions for children with obesity.

All guidelines selected have been developed and published by important national paediatric societies or National Institutes for Health. The guidelines are structured with an initial section detailing their development process, the underlying scientific evidence, and their proven efficacy. This section is followed by a list of recommendations addressing the identification, classification, assessment, and nutritional management of obesity. The majority of these guidelines pertain to child populations aged 0–18 years. However, two specific guidelines, Spanish Ministry for Health (SPAIN) and NICE, encompass children aged 6–18 and 2–18 years, respectively. Table 4.1 shows distinct attributes and abbreviated forms of selected European guidelines.

The recommendations provided by all the guidelines are established upon the most robust research evidence available and prior published guidelines. The level of evidence and strength of recommendation are determined in consensus with the respective working groups. Each working group has prepared an initial report for each specific recommendation, followed by a concise description of the supporting scientific evidence, epidemiological data, and pertinent notes. The outcomes of every study, whether observational or based on trials, have been assessed by each group using the international Grading of Recommendations Assessment, Development, and Evaluation (GRADE) method (http://www.gradeworkinggroup.org/). The guidelines are primarily aimed at paediatricians, general practitioners, and healthcare professionals working in primary care settings, alongside specialists in fields like nutrition and dietetics, who operate within specialised care. It is important to emphasise that these recommendations cannot supplant clinical judgment, as they cannot be universally applicable due to the wide range of clinical scenarios. Therefore, the establishment of clinical discretion remains of utmost importance in the application of the guidelines (Woolf et al. 1999). All the chosen guidelines can be accessed online.

4.2.1 Obesity Definition

The detection criteria and references for the definition of obesity in children are presented in Table 4.2. Body mass index (BMI) percentiles are recommended as a routine measure for the diagnosis of obesity and to define the weight status of children in all guidelines, except for EASO–European Federation of the Association of Dietitians (EFAD) guidelines. Furthermore, all guidelines advocate for healthcare providers such as paediatricians, family doctors, and primary care professionals to rely on standard growth charts when assessing children's weight status.

The definition of obesity varies among guidelines, resulting in four sets of cut-off criteria. The first set, defined by the National Board of Health and the Danish College of General Practitioners, identifies obesity as exceeding the 99th percentile of BMI (DSAM n.d.; Tinggaard et al. 2014) and the Danish guidelines (DPS) adopted this

Table 4.1 Features of selected guidelines

Title	Author/organisation	Country	Population age group of children	Publication year
Clinical practice guideline for the prevention and treatment of childhood and juvenile obesity	Spanish Ministry for Health (SPAIN)	Spain	6–18 years	2009
Management of obesity —a national clinical guideline	Scottish Intercollegiate Guidelines Network (SIGN)	Scotland	0–18 years	2010
Identification, assessment and management of overweight and obesity in children, young people and adults	National Institute for Health and Care Excellence (NICE) (National Institute for Clinical Excellence)	United Kingdom	2–18 years	2014
Danish clinical guidelines for examination and treatment of overweight and obese children and adolescents in a pediatric setting	Johansen et al./Danish Pediatric Society (DPS)	Denmark	Not specified	2015
Diagnosis, treatment and prevention of pediatric obesity: consensus position statement of the Italian Society for Pediatric Endocrinology and Diabetology and the Italian Society of Pediatrics	Valerio et al./Italian Society of Paediatrics and the Italian Society for Paediatric Endocrinology and Diabetology (SIP-SIEDP)	Italy	0–18 years	2018
EASO and EFAD position statement on medical nutrition therapy for the management of overweight and obesity in children and adolescents	Hassapidou et al./European Association for the Study of Obesity (EASO) and European Federation of the Association of Dietitians (EFAD)	Europe	Not specified	2023

definition. The second set of criteria suggests an alternative cut-off (obesity = 98th percentile of) (Hernández et al. 1988; Cole 1997), and both the SPAIN and Scottish Intercollegiate Guidelines Network (SIGN) guidelines adhered to this threshold. The third set, provided by the NICE guidelines, recommends two different cut-offs: one for clinical practice (obesity = ≥91st percentile of BMI) (Cole 1997) and another for assessing comorbidity (obesity = ≥98th percentile of BMI) (Cole 1997). Finally, the fourth set of cut-offs, outlined by WHO in 2006 and 2007 (De Onis 2006; De Onis

Table 4.2 Recommendations and references for obesity definition in the included guidelines

Obesity detection criteria	Reference	Guideline
Index: BMI percentiles Obesity >99th percentile	National Board of Health and the Danish College of General Practitioners (DSAM); Danish growth references for boys and girls (Tinggaard et al. 2014)	DPS (Johansen et al. 2015)
Index: BMI percentiles Obesity ≥98th percentile	1990 UK reference data (UK90) (Cole 1997)	SIGN (Scottish Intercollegiate Guidelines Network 2010)
Index: BMI percentiles Obesity ≥98th percentile	Growth curves and tables of Hernández et al. (1988)	SPAIN (Spanish Ministry for Health 2009)
Indices: weight-to-length ratio for 0–2 years; BMI percentiles for 2–5 years, and 5–18 years For 0–2 and 2–5 years: Obesity ≥99th percentile For 5–18 years: obesity >97th percentile; severe obesity ≥99th percentile	WHO (2006, 2007), De Onis (2006), De Onis et al. (2007)	SIP-SIEDP (Valerio et al. 2018)
Index: BMI percentiles Tailored clinical intervention recommended with a BMI ≥91st percentile; assessment of comorbidity recommended with BMI ≥98th percentile	1990 UK reference data (UK90) (Cole 1997)	NICE (National Institute for Clinical Excellence 2014)
Unspecified	Unspecified	EASO-EFAD (Hassapidou et al. 2023)

et al. 2007), is followed by the Italian Society of Paediatrics and the Italian Society for Paediatric Endocrinology and Diabetology (SIP-SIEDP) guidelines, incorporating different thresholds for various age groups.

4.2.2 Recommendations for Nutritional Management and Dietary Intervention

Recommendations regarding nutritional assessment, dietary intervention, monitoring and evaluation of the intervention, the levels of evidence (LOE), and the degree of each recommendation are summarised in Table 4.3.

4.2.2.1 Nutritional Assessment

Many of the selected guidelines recommend a targeted evaluation of the nutritional status of children dealing with obesity. This assessment should encompass factors like individual lifestyle and behaviour, psychosocial stress, and other psychological

Table 4.3 Different guidelines propose varied approaches to nutritional assessment, dietary intervention, monitoring and evaluation, and dietary recommendations for the management of obesity in children

Topic	DPS (Johansen et al. 2015)	SIGN (Scottish Intercollegiate Guidelines Network 2010)	SIP-SIEDP (Valerio et al. 2018)	SPAIN (Spanish Ministry for Health 2009)	NICE (National Institute for Clinical Excellence 2014)	EASO-EFAD (Hassapidou et al. 2023)
Nutritional assessment						
Food/nutrition	Assess history of diet[a]	Assess history of diet (LOE II-C) Assess food quality (LOE II-C)	Assess eating behaviour (LOE I-A) Assess meal structure and distribution (LOE I-A) Assess food quality and quantity (including portion sizes) (LOE I-A)	Assess eating behaviour (LOE VI-D) Assess meal structure and distribution (LOE VI-D) Assess number of snacks or drinks between meals and their composition (LOE VI-D)	Explore eating patterns[a]	Not included
Anthropometry	Calculate BMI[a] Analyse growth curves[a] Waist circumference[a]	Calculate BMI (LOE II-C) Analyse growth curves (LOE II-C)	Calculate BMI (LOE V-A) Analyse growth curves (LOE V-A)	Calculate BMI (LOE VI-D) Analyse growth curves (LOE VI-D)	Calculate BMI[a] Analyse growth curves[a] Waist circumference[a]	Not included
Biochemical information, instrumental tests, and procedures	Measure blood pressure[a] Analyse blood glucose[a] Analyse blood lipids[a] Assess liver	Measure blood pressure (LOE I-D) Measure blood glucose (LOE VI-D) Analyse blood lipids (LOE II-D)	Measure blood pressure (LOE III-A) Measure blood glucose (LOE V-A) Interpret lipid profile (LOE I-A)	Measure blood pressure (LOE VI-D) Measure blood glucose (LOE VI-D)	Measure blood pressure[a] Assess lipid profile Measure blood glucose [a] Measure HbA[a]	Not included

	function tests[a] Assess neurological test[a] Assess vitamin D deficiency[a] Use DEXA scan		Assess transaminases and liver function (LOE V-B)	Check lipid profile (LOE VI-D)		
Nutrition-centred physical findings	Assess sleep[a] Assess the presence of acanthosis nigricans[a]	Not included	Not included	Not included	Not included[a]	Not included
Physical activity	Assess physical activity[a]	Assess physical activity (LOE II-C)	Assess physical activity (LOE I-B)	Assess physical exercise habits (LOE I-B)	Explore physical activity levels[a]	Not included
Patient history	Assess comorbidities (such as hypertension, dyslipidaemia, insulin resistance, prediabetes, type 2 diabetes, metabolic syndrome, and cardiovascular disease)[a] Assess social history[a] Assess psychosocial history[a]	Assess comorbidities (such as hypertension, dyslipidaemia, insulin resistance, type 2 diabetes, and cardiovascular disease) (LOE II-C) Assess psychosocial distress (LOE II-D) Assess socio-economic factors (LOE V-C)	Assess comorbidities (such as hypertension, dyslipidaemia, insulin resistance, type 2 diabetes, and cardiovascular disease) (LOE I-A) Recognition of psychosocial correlates (LOE V-A)	Assess comorbidities (such as hypertension, dyslipidaemia, insulin resistance, and type 2 diabetes) (LOE VI-D) Assess family history (LOE VI-D) Assess social history (LOE VI-D)	Assess comorbidities (such as hypertension, hyperinsulinemia, dyslipidaemia, and type 2 diabetes)[a] Assess psychosocial distress[a] Assess family history[a] Assess socio-economic factors[a]	Not included

(continued)

Table 4.3 (continued)

Topic	DPS (Johansen et al. 2015)	SIGN (Scottish Intercollegiate Guidelines Network 2010)	SIP-SIEDP (Valerio et al. 2018)	SPAIN (Spanish Ministry for Health 2009)	NICE (National Institute for Clinical Excellence 2014)	EASO-EFAD (Hassapidou et al. 2023)
Eating disorders	Not included	Not included	Assess the presence of binge eating disorders (LOE V-B)	Screen for eating disorders (LOE VI-D)	Not included	Not included
Nutritional intervention						
Meal	Structure eating with 4–5 (6) meals a day (DNI)[a] Promote consuming breakfast every morning[a] Consume snacks between main meals (whole grain products, vegetables, cold cuts low-fat)[a]	Not included	Structure eating with 5 meals a day (3 meals and no more than 2 snacks) (LOE V-B) Have an adequate breakfast (LOE II-B) Avoid eating between meals (LOE III-B)	Divide the daily food intake into 4–5 meals per day (LOE I- A)	Not included	Not included
Food	Promote consuming vegetables/salads in large amounts[a], maximum 2 fruit servings a day, water and half a litre of skimmed milk a day[a] Avoid soft drinks	Promote consuming at least 5 portions of fruit and vegetables a day (LOE I-D) Limit salt intake (LOE I-D) Recommend choosing lower-fat	Increase the intake of fruits, vegetables, and cereals rich in fibre (LOE VI-A) Avoid high-energy-density and low-nutritional-density foods (fast-foods,	Balance diet and eating habits (LOE I-A) Promote consuming high amounts of cereals, 5 portions of fruits and vegetables daily, milk and	Balance diet and eating habits[a] Promote consumption of a variety of fruits and vegetables to at least 5 portions per day[a] Promote	Promote the consumption of nutrient-rich foods with lower energy density, especially vegetables and fruit (LOE III-A) Promote the reduction of

	consumption Limit candies and other sweets consumption[a] Avoid high–energy-density and low-nutritional-density foods (fast foods)[a] Avoid butter or spread on the bread[a]	foods (LOE II-D) Limit intake of energy-dense foods (LOE II-B)	take-away foods) (LOE III-B) Avoid sugar-sweetened beverages and juices (LOE III-B)	dairy products daily, fish 2 to 4 times per week, small amounts of fats and foods rich in fats, and water (LOE III-D) Limit salt intake (LOE II-D) Limit sugary drinks (LOE II-B)	consumption of starchy foods such as whole grain bread, pasta, and rice[a] Promote consumption of low fat (especially saturated fat), sugar, and salted foods[a]	energy-dense nutrient-poor beverages (LOE IV-A) Promote the reduction of highly processed or energy-dense meat (LOE III-A) Promote the reduction of fat dairy foods (LOE III-A)
Portion sizes	Limit portions[a]	Not included	Portions must be appropriate for the individual's age, gender, current weight, and activity level (LOE I-A)	Avoid copious portions (LOE I-D)	Not included	Not included
Diet	Prescribe energy-reduced diet[a]	Healthier eating and decreasing energy intake (LOE I-D)	Promote and support a balanced and varied diet (LOE I-A) Prescribe a hypocaloric diet based on sex, age, and ideal weight for stature (LOE VI-A) Prescribe very low caloric diet for patients with severe	Recommend a healthy, balanced diet according to the healthy eating pyramid (LOE I-A) Calorie intake must be lower than the energy children expend, and changes to their diet must be customised and consistent with	Dietary changes should be individualised, sustainable, age appropriate and consistent with healthy eating advice[a] Prescribe energy-reduced diet[a] Do not use unduly restrictive and	Healthier eating and decreasing energy intake (LOE III-A) Prescribe diet with moderate energy intake deficit (LOE V-A)

(continued)

Table 4.3 (continued)

Topic	DPS (Johansen et al. 2015)	SIGN (Scottish Intercollegiate Guidelines Network 2010)	SIP-SIEDP (Valerio et al. 2018)	SPAIN (Spanish Ministry for Health 2009)	NICE (National Institute for Clinical Excellence 2014)	EASO-EFAD (Hassapidou et al. 2023)
			obesity (LOE III-C) Prescribe the "traffic light diet" (LOE III-C) Prescribe hypocaloric diets with low glycaemic index and low glycaemic load (LOE I-C)	healthy eating (LOE I-A) Do not use restrictive, unbalanced diets (LOE I-D)	nutritionally unbalanced diets[a]	
Meal environment	Avoid screen during meal (TV/PC/ PlayStation)[a]	Not included	Eat as family (LOE I-B)	Not included	Not included	Not included
Parental involvement	Families should be included in the treatment[a]	Family, at least 1 parent/carer, and the child should be included in the treatment (LOE II-D)	Families should be included in the treatment (LOE II-A)	Families should be included in the treatment (LOE I-D)	Families should be included in the treatment[a]	Families should be included in the treatment (LOE VI-A)
Lifestyle intervention	Multifactorial treatment including diet, physical activity, and behaviour management[a]	Multifactorial treatment including diet, increased physical activity, and reduction of sedentary behaviour (LOE I-B)	Multifactorial treatment including diet, physical activity, and sedentariness management (LOE III-A)	Combined interventions including diet, physical exercise, and changes to behaviour, with family involvement (LOE I-A)	Multicomponent treatment including diet, physical activity, and behaviour management[a]	Multicomponent treatment including diet, physical activity, and behaviour management (LOE VII-A)

Physical activity	At least 1 hour of exercise every day—every session should last at least 20 minutes[a]	At least 60 minutes of moderate activity per day (LOE IV-D)	Promote physical exercise in association with diet (LOE I-A) Combine aerobic and resistance exercises for at least 60 minutes every day (LOE I-B)	Increase spontaneous physical activity by more than 1 hour a day (outdoor play, helping with housework, walking to school, using stairs instead of lifts (LOE I-A). Recommend physical exercise suited to age and interests. It should begin slowly and gradually increase in intensity and duration (LOE I-A)	Encourage children to increase their level of physical activity, even if they do not lose weight as a result[a] Encourage children to do at least 60 minutes of moderate or greater intensity physical activity each day in sessions lasting 10 min or more[a] Make the choice of activity with the child, and ensure it is appropriate to the child's ability and confidence[a]	Promote physical activity (LOE IV-A)
Sedentariness	Reduce non-academic screen time to max 2 hours daily[a]	No more than 14 hours screen time per week (LOE II-D)	Reduce the time spent in sedentary behaviour (watching television, video gaming, and Internet surfing) (LOE I-B) Use of active video games may be suggested to increase daily energy (LOE I-B)	Reduce sedentary lifestyle (LOE I-D) A maximum of 1½ hours a day watching television or playing video games (LOE I-B) Televisions, video consoles, and computers should be removed from the bedrooms (LOE I-A)	Encourage children to reduce inactive behaviours, such as sitting and watching television, using a computer, or playing video games[a]	Reduce the time spent in sedentary behaviour (LOE III-A)

(continued)

Table 4.3 (continued)

Topic	DPS (Johansen et al. 2015)	SIGN (Scottish Intercollegiate Guidelines Network 2010)	SIP-SIEDP (Valerio et al. 2018)	SPAIN (Spanish Ministry for Health 2009)	NICE (National Institute for Clinical Excellence 2014)	EASO-EFAD (Hassapidou et al. 2023)
Care coordination and implication	Paediatricians[a] and dietician[a]	Dietician and paediatric specialist (LOE I-D)	Paediatricians (primary care) (LOE III-A) District or hospital outpatient services (secondary care) (LOE VI-A) Specialised Centre for paediatric obesity (third level care) (LOE VI-A)	Healthcare professionals (LOE I-A) Specialist outpatient centres (LOE I-D)	Paediatricians[a] and dietician[a]	Dietician and healthcare professionals (LOE IX-A)
Nutritional monitoring and evaluation						
Follow-up	Not included	Intervention both intensive and long-lasting (≥6 months) (LOE I-B)	Intervention 6–12 months	Not included	Offer regular, non-discriminatory long-term follow-up[a]	Not included
Monitoring indicators	Not included	Annual monitoring of BMI centile (LOE I-B)	Monitor the BMI z-score (LOE V-B)	Monitoring of BMI centile (LOE I-D) and other behavioural indicators (LOE VI-B)	Monitoring of BMI centile[a]	Not included

[a]Level of evidence (LOE) and grade of recommendation were not reported in the guidelines. The level of all recommendations was considered acceptable in general

aspects. This approach aims to support the development of tailored treatment strategies and programmes. However, it is important to note that the nature and level of recommendations vary among the guidelines, as detailed in Table 4.3. Notably, the EASO-EFAD guidelines do not provide any assessment recommendations. Regarding the physical examination, all guidelines uniformly suggest using BMI as a measure of overall adiposity. Additionally, NICE and DPS guidelines specifically recommend measuring waist circumference to provide additional insights into the risk of potential long-term health issues (Bassali et al. 2010). Concerning biochemical and instrumental evaluation, the measurement of blood pressure, fasting blood glucose, and lipid profile is recommended by guidelines according to the most up-to-date and shared reference criteria (National High Blood Pressure Education Program Working Group on High Blood Pressure in Children and Adolescents 2004; De Jesus 2011; Friedemann et al. 2012; Peterson and McBride 2012; Lurbe et al. 2016; Di Bonito et al. 2017). Some differences in age and measurement criteria are reported in the SIP-SIEDP guidelines. According to these guidelines, it is advised to assess blood pressure in children with obesity starting from the age of 3 years. Additionally, the guidelines recommend considering blood pressure measurements in children under 3 years old only in the presence of cardiac malformations, genetic diseases, or acquired or congenital kidney diseases. Furthermore, the determination of elevated blood pressure levels should be based on percentiles (sex- and age-specific) of systolic and diastolic blood pressure relative to height percentiles, following both the SIP-SIEDP and SPAIN guidelines (National High Blood Pressure Education Program Working Group on High Blood Pressure in Children and Adolescents 2004; Lurbe et al. 2016). Fasting blood glucose measurement as the first step in screening for prediabetes and type 2 diabetes is recommended in all children with obesity from the age of 6 years in the SIP-SIEDP guidelines. Evidence from the literature indicates that prediabetes is detectable in approximately 5% of children under the age of 10 who are obese (Di Bonito et al. 2017). The measurement of cholesterol, HDL-cholesterol, and triglycerides is recommended in all children with obesity from the age of 6 years in the SIP-SIEDP guidelines. The same guidelines indicate that the diagnosis of dyslipidaemia must be made based on national reference values and must be repeated after 3 years or more frequently if the development of other cardiometabolic comorbidities occurs (Peterson and McBride 2012). Liver function tests, including the assessment of alanine transaminase, alkaline phosphatase, bilirubin, and gamma-glutamyl transferase, are only suggested in the DPS and SIP-SIEDP guidelines. The NICE guidelines advise providing children and their families with information regarding the purpose of all tests, the methodology used for these tests, and the outcomes or results obtained.

All guidelines emphasise the importance of evaluating a child's dietary habits as a crucial starting point for nutritional intervention. Both the SPAIN and SIP-SIEDP guidelines recommend gathering information on meal distribution and structure, as well as assessing the quality and quantity of foods consumed, along with food preferences or aversions. The DPS, SPAIN, and SIP-SIEDP guidelines propose the use of a food diary as a valuable tool for assessing eating behaviours (Burrows et al. 2010; Burke et al. 2011). This diary should be completed by children with one of their parents and subsequently evaluated by specialised operators (Burrows et al. 2010; Burke et al. 2011). In addition, the assessment of psychosocial or emotional distress, including the incidence of bullying, teasing, solitude, and low self-esteem, is suggested in four guidelines (DPS, SIGN, NICE, and SIP-SIEDP). This evaluation is regarded as essential for advancing targeted strategies aimed at enhancing the outcomes of dietary interventions (Anderson et al. 2007; Verrotti et al. 2013).

Evaluation of sleep quality, including apnoea and the sleep duration, and the assessment of the presence of acanthosis nigricans, especially in the neck, armpits, and groin (often associated with hyperinsulinemia) is recommended only by the DPS guidelines. Several pieces of evidence indicate that insufficient sleep duration can lead to significant alterations in body mass index z-score (Fatima et al. 2015; Derks et al. 2017). Only the NICE and SIP-SIEDP guidelines suggest assessing the willingness of the child and family to make lifestyle changes. It is noteworthy that none of the guidelines offers recommendations for evaluating the prescription of obesogenic drugs.

4.2.2.2 Dietary Intervention

In each set of guidelines, dietary changes should be a key component of a comprehensive lifestyle programme, which includes parental or family involvement, increased physical activity, and reduced sedentary behaviour. The supporting evidence for this recommendation indicates that lifestyle interventions integrating dietary changes result in a reduction in at least one indicator of adiposity when compared to control groups without dietary interventions (Duncanson et al. 2017). Furthermore, the combination of dietary adjustments and increased physical activity is a superior strategy for reducing adiposity compared to interventions that exclusively focus on physical activity (Burrows et al. 2011; DeBar et al. 2012). There is a unanimous consensus in all guidelines in emphasising that nutritional treatment should promote lasting changes in a child's dietary habits and lifestyle, aiming for a gradual reduction in BMI rather than pursuing rapid weight loss. Overall, the compilation of a diet history questionnaire is considered an essential educational tool, serving to improve children's awareness of their current habits and initiating the intervention process. The NICE guidelines particularly recommend the importance of personalised interventions that take into account both the child's and the family's preferences and needs. The success of the treatment is deemed entirely dependent on the active involvement of the children's parents or family. Several pieces of evidence support the inclusion of parents to improve the results of nutritional treatment in children aged 3–12 years (Tomayko et al. 2016; Ells et al. 2018). An intervention study that involved the entire family in the treatment of obese children between the ages of 4 and 6 showed a significant difference in adiposity when compared to a child-only treatment (Ells et al. 2018). Nevertheless, the guidelines lack specific recommendations regarding the nature, intensity, and preferred approach to involving parents. Across all the selected guidelines, it is consistently emphasised, supported by substantial evidence, that nutritional interventions should be administered by specialised healthcare professionals, such as paediatricians, dietitians, and other relevant healthcare professionals (Looney and Raynor 2014; Larsen et al. 2015, 2016; Small et al. 2015). A systematic review showed that 67% of the dietitian-administered dietary interventions determine a significant decrease in total energy intake compared to interventions administered by non-dietitian professionals (Duncanson et al. 2021).

Regarding dietetic approaches, most guidelines recommend a balanced, varied, and age-appropriate diet. The NICE guidelines specify that the diet should also be sustainable. This diet should align with the recommendations of age- and gender-specific micronutrients and the indications of healthy eating guidelines, which may vary among countries (Food and Agriculture Organization of the United Nations 2022). In this context, the guidelines provide diverse dietary recommendations related to meals, food choices, portion sizes, and meal settings, which are summarised in Table 4.3. Among these recommendations, the most frequent advice is to increase the consumption of fruits and vegetables, while reducing the intake of high-calorie, low-nutrient-density foods, such as sweetened beverages. There is a substantial body of evidence supporting these recommendations. Some studies showed that an increase in the intake of fruits and vegetables determines a significant reduction in BMI z-scores, after 4 weeks, in children and adolescents with obesity and hypercholesterolemia (Macknin et al. 2015). Several intervention studies have reported improvements in adiposity outcomes as a result of the limitation of high-energy and low-nutrient-density food intake in children (Ebbeling et al. 2012; Epstein et al. 2015). DPS guidelines offer specific recommendations on meal composition. According to these guidelines, breakfast should consist of whole grain products (bread) with lean meats, oatmeal, whole grains, eggs, and low-fat milk. Lunch should include wholemeal bread with cold cuts and lean vegetables. Snacks should consist of whole grain products (bread, oatmeal, etc.), vegetables, and low-fat meats (ideally containing 6% fat, equivalent to 6 g of fat per 100 g of the product). For dinner, meals should be served in limited portions according to model T or the "healthy plate model" (Johansen et al. 2015).

Some guidelines, including DPS, SPAIN, NICE, and EASO-EFAD, recommend a dietary model based on the reduction of total energy intake. This dietary model should be balanced, age-appropriate, and consistent with healthy eating guidelines (Food and Agriculture Organization of the United Nations 2022). According to the NICE guidelines, to obtain a moderate reduction in total energy intake, it is suggested to keep fat intake at 35% of food energy, reduce saturated fats and added sugars to 11% of food energy, and increase fibre intake by 18 g per day.

A low-calorie diet is not recommended by SIP-SIEDP guidelines since it is deemed ineffective in the medium to long term and may result in relapses and failures, increasing the risk of dropout and progression of obesity into more complicated forms (Astrup et al. 2015). However, the same guidelines specify that, if necessary, a hypocaloric diet should include 1 g/kg/day of protein, 45–60% of total calories in carbohydrates, a quantity of simple sugars <15% of total calories, a quantity of lipids equal to 20–35% of total calories (from 4 years of age), and a quantity of saturated fatty acids <10% of total calories. The diet should be tailored based on sex, age, and ideal weight for stature. The SIGN guidelines state that there is no evidence to suggest any particular dietary or macronutrient manipulation.

The SIP-SIEDP guidelines recommend a traffic-light diet that combines foods into three categories: green (low-energy, high-nutrient foods to be consumed often), yellow (moderate-energy foods to be consumed in moderation), and red (high-energy, low-nutrient foods to be consumed sparingly) for children aged 8–12 years with obesity. Evidence has found this diet can determine a significant improvement in BMI in children aged 8–12 years even in the long term (Epstein et al. 2008). The SPAIN and NICE guidelines do not recommend the use of restrictive, unbalanced diets.

Of note, only a few guidelines recommend intensive dietary approaches such as the adoption of a very low-energy diet. Specifically, the SIP-SIEDP and EASO-EFAD guidelines recommend a very low-calorie diet consisting of 600–800 kcal/day for patients with severe obesity. This approach requires close medical monitoring and is ideally implemented in specialised paediatric centres. The duration of this restrictive diet must not exceed 10 weeks. A systematic review that evaluated very low-energy diet interventions (≤800 kcal/day) and a subsequent meta-analysis both reported a substantial reduction in BMI, approximately 3.4 kg/m^2, during intervention periods spanning from 5 to 14.5 months (Andela et al. 2019; Duncanson et al. 2021). Also, another meta-analysis provided evidence indicating significant improvements in cardiometabolic outcomes, including blood pressure, total cholesterol, blood glucose, and insulin levels (Andela et al. 2019). There is no evidence on the medium-term effectiveness of diets with low glycaemic index and low glycaemic load compared to other dietary approaches, according to the SIP-SIEDP guidelines (Esfahani et al. 2011; Kirk et al. 2012). On the contrary, the EASO-EFAD guidelines highlight that in the context of broader lifestyle interventions, low glycaemic load interventions are as effective as interventions altering macronutrient proportions of a hypocaloric diet (Mirza et al. 2013; De Ferranti et al. 2015). Overall, the selected guidelines provide limited guidance on age- and gender-specific dietary approaches for managing obesity, especially severe obesity.

4.2.2.3 Physical Activity

All the guidelines included in the present report recommend incorporating physical activity alongside dietary changes as part of the nutritional treatment for children with obesity. However, the specifics provided within these guidelines vary. Supporting this recommendation, various pieces of evidence indicated that the combination of exercise and dietary modifications leads to improvements in body composition and a reduction in cardiometabolic risk factors (Atlantis et al. 2006; McGovern et al. 2008; Janssen and LeBlanc 2010; Brambilla et al. 2011; Kelley et al. 2015). The NICE and SPAIN guidelines recommend tailoring physical activity plans to suit a child's age, interests, and abilities, aiming to foster and sustain their engagement. The NICE guidelines highlight the benefits of encouraging physical activity regardless of BMI changes. In line with the recommendations of the Word Health Organization (Chaput et al. 2020), the NICE, SIGN, SPAIN, and SIP-SIEDP guidelines recommend that children engage in at least 60 min of moderate to vigorous activity per day (Strong et al. 2005; Janssen 2007). The NICE guidelines propose that the recommended 60 min of activity can be divided into shorter sessions of at least 10 min each, distributed throughout the day. Limited evidence indicates that engaging in higher-intensity exercise is more effective in altering body composition (Atlantis et al. 2006). Several guidelines, including those of SIGN, NICE, and SPAIN, encourage and provide support for children to incorporate more spontaneous physical activity into their daily routine, such as active play, walking, and taking the stairs.

4.2.2.4 Sedentariness

All the selected guidelines recommend encouraging children to reduce sedentary activities, such as watching TV, using computers, and playing video games. Evidence in support of this recommendation shows a strong relationship between TV hours, obesity, and cardiometabolic risk factors, which could potentially be attributed to overfeeding (LeBlanc et al. 2012; Te Velde et al. 2012; De Rezende et al. 2014). Conversely, there is a modest correlation suggested between sedentary behaviour, obesity, and cardiometabolic risk factors (Pearson and Biddle 2011). The DPS, SIP-SIEDP, SIGN, and SPAIN guidelines recommend limiting non-educational screen time, such as watching television and playing computer games, to a maximum of 1.5–2 hours per day, which is likely to encourage increased physical activity. The SPAIN guidelines also recommend removing televisions, video consoles, and computers from bedrooms. Lastly, the SIP-SIEDP guidelines recommend the use of active video games as a means to boost daily energy expenditure (Gao and Chen 2014; McNarry and Mackintosh 2016).

4.2.2.5 Nutritional Monitoring and Evaluation

Regarding the duration of the follow-up, the SIGN, SIP-SIEDP, and NICE guidelines recommend a nutritional intervention lasting at least 6 months to 1 year. This recommendation is supported by different pieces of evidence that show the limited efficacy of diet- and lifestyle-based treatment in achieving long-term BMI reduction (McGovern et al. 2008; Oude Luttikhuis et al. 2009).

Most guidelines recommend using the BMI standard deviation score (BMI-SDS) to estimate weight loss. The reduction of the BMI-SDS is considered the best indicator of weight loss, accounting for the age and sex of the children during interventions (Bryant et al. 2014). Since the potential percentage of weight loss that can be achieved is typically modest, relying solely on BMI-SDS could lead to a perception of failure in both the family and the children, according to the SIP-SIEDP guidelines. Therefore, these guidelines recommend incorporating other behavioural indicators related to diet, lifestyle, physical fitness, and quality of life into the assessment process. This more comprehensive approach aims to ensure better adherence to treatment by both the children and their families (Finne et al. 2013; Kolotourou et al. 2013).

As is evident, Europe lacks a harmonised approach to tackling obesity through diet management. The development of overarching guidelines applicable across all European nations could substantially mitigate regional variations and provide more effective guidance for dietary practices.

4.3 Prospects: Personalised Strategies

Despite ample scientific evidence demonstrating the efficacy of interventions in line with clinical practice guidelines, the prevalence of childhood obesity continues to be substantial in various European regions, as noted by WHO. It has become clear that a "one-size-fits-all" approach is insufficient when it comes to developing prevention and treatment programmes (Motevalli et al. 2021). Recent evidence has revealed the influence of genetics, as well as gut microbiota composition, in tailoring personalised diets for individuals with obesity and diabetes (Zeinalian et al. 2022). Also, personalised gut-microbiota-targeted diets have proven effective in improving responses to therapeutic treatments (Chen et al. 2022; Zeinalian et al. 2022). The recent development of personalised dietary approaches represents a promising research field that runs in parallel with the established domain of precision medicine (Holzapfel and Drabsch 2019; Barrea et al. 2020; Chen et al. 2022). Both fields attempt to prevent and treat diseases in a targeted way, taking into account personal characteristics (Chen et al. 2022; Zeinalian et al. 2022). Personalised nutrition that considers the interactions among genetics, environmental and personal characteristics (such as eating habits, life behaviours, levels of physical activity, and sleep hours), and the population of microbes that live in our gut (the so-called "gut microbiota") may be essential to provide tailored and more accurate nutritional advice for a healthy diet (Bray et al. 2016; Gibson and Sainsbury 2017; Koliaki et al. 2018). In addition, according to the general consensus, the implementation of a well-proposed personalised programme not only maximises desirable outcomes but also contributes to the sustainable respect of a healthy lifestyle model (Koliaki et al. 2018). A valuable example of this type of approach is embodied by the Stance4Health project (Dello Russo et al. 2022). This project represented an innovative and effective way of developing a complete Smart Personalised Nutrition (SPN) service based on the use of mobile technologies, as well as tailored food production able to optimise gut microbiota activity and long-term consumer engagement (Dello Russo et al. 2022). For this purpose, the nutritional i-Diet Stance4Health app was implemented with a very complex algorithm capable of creating weekly menus tailored to individual people and the bacteria that cohabit in their intestines, also considering the beneficial effect of specific natural extracts (Dello Russo et al. 2022). In the current literature, evidence that supports personalised nutrition for specific treatment of obesity in children and adolescents comes from several studies. In a recent review, Motevalli et al. have shown how the personalisation of nutritional programmes (e.g. meal planning, portion control, and selection of healthy snacks), physical activity programmes (using technological supports), and in general the modulation of the recommendations on the personal needs of the child and the family have been effective in modifying the behaviours of obese children and adolescents (Motevalli et al. 2021). In particular, this strategy of personalised intervention, aetiology-dependent (EPISTCO), stressed the importance of incorporating the different approaches related to a healthy lifestyle (nutritional, physical activity, and quality of sleep) and adapting them to the individual needs and characteristics of people to ensure therapeutic personalisation (Motevalli et al.

2021). Favourable results in moderate–severely obese children were also reported in a review study that evaluated the effects of educational programmes based on the personalised nutritional approach, Nutrition Care Process (NCP), comprising education on meal planning, portions, selection of healthy snacks, and foods of vegetable origin (Kim et al. 2019). In addition to these studies of nutritional interventions, other studies have highlighted the effectiveness of other personalised interventions on weight or other health outcomes in children and adolescents. Ghanvatkar et al. demonstrated that interventions aimed at increasing physical activity are more effective at modifying health behaviours when personalised and technology-based, as opposed to relying on a "one-size-fits-all" generic advice (Ghanvatkar et al. 2019). A study in obese or diabetic adolescents highlighted that performing personalised exercises could improve physical activity levels and personal health when associated with family support (Faulkner et al. 2014). Evidence also indicated that children's health behaviours, in particular sleep habits, could be positively modified through personalised interventions on mothers (Kouloughioti et al. 2013). In light of this, the future effort should be to develop ever more precise algorithms capable of predicting individual food responses, based on specific parameters that include nutritional indices and lifestyle (physical activity, sleep, stress) to achieve maximum personalisation of the therapeutic intervention and obtain the best results, in terms of both prevention and treatment of chronic non-communicable diseases (Dello Russo et al. 2022).

4.4 Gut Microbiota and Personalised Nutrition

The gut microbiota is a thriving ecosystem of microorganisms that colonise the human intestine (Singh et al. 2017), whose activity and composition are influenced by genetic and environmental factors (de la Cuesta-Zuluaga et al. 2019). The gut microbiota plays a significant role in the pathogenesis, progression, and treatment of many diseases, including metabolic conditions (Cho and Blaser 2012; Le Chatelier et al. 2013; Parekh et al. 2014). Several studies have shown that the alteration of the Firmicutes/Bacteroidetes ratio can contribute to the development of central obesity and type 2 diabetes mellitus (Noce et al. 2019). Although the early stages of life and the genetic characteristics of the host have considerable effects on the microbiota, this community retains a degree of flexibility and can be modulated through exposure to various environmental factors, particularly the diet (David et al. 2014; Leeming et al. 2019). The microorganisms comprising the intestinal microbiota respond differently and are species-specific to the type and amount of food. Furthermore, the effects of a particular diet can be influenced by a combination of host and microbiome characteristics (Zeevi et al. 2015; Rothschild et al. 2018). In light of this, personalised nutrition is increasingly recognised as a new therapeutic pathway that can modulate host–microbiota interactions to prevent and control metabolic disorders (Sarfraz et al. 2022). The main challenge of personalised nutritional approaches is to determine the main characteristics of the gut microbiome that could help in the prediction of the metabolic response of the host to food components. This information can then inform the design of a personalised diet to yield favourable results (Sarfraz et al. 2022). In this approach, it is therefore very important to know how the host, microbiome, and food exposure work together and affect each other. All dietary macronutrients act on the intestinal microbiota, causing a change in its composition with positive or negative effects on the health of the host. Several studies of human interventions have shown that these changes can occur quickly and reproducibly (David et al. 2014). Specifically, consuming a diet entirely based on food of animal or plant origin causes changes in the intestinal microbiota in 4 days, while the change in the consumption of the type of fat or fibre manifests itself within 2 weeks (Tomova et al. 2019; Losno et al. 2021). Conversely, minor modifications in some nutritional components do not easily disrupt the resilience of the intestinal microbiota (Korem et al. 2017). Several studies have shown that a diet rich in animal proteins, both in the long and short terms, consistently increases the level of *Bacteroides* (Wu et al. 2011) while decreasing the abundance of saccharolytic microorganisms (David et al. 2014). In contrast, the consumption of a plant protein diet significantly increases the levels of *Lactobacillus* and *Bifidobacterium species* (Dominika et al. 2011). A positive role in maintaining a thriving microbial ecosystem is covered by dietary fibre derived from fruits, vegetables, and whole grains such as wheat germ (Healey et al. 2018). Dietary fibres interact directly with intestinal microbes, favouring the production of the short-chain fatty acids (SCFA) that increase the immune defences of organisms (Cronin et al. 2021). A randomised controlled study showed an increase in microbial diversity and in SCFA and a decrease in Enterobacteriaceae after the consumption of whole grains compared to the consumption of refined wheat (in the control group) (Vanegas et al. 2017). A high intake of dietary fat, in particular of saturated fatty acids, seems to favour the development of a pro-inflammatory intestinal microbiota, resulting in increased intestinal permeability and, consequently circulating levels of lipopolysaccharides (LPS), as well as the development of chronic degenerative diseases such as obesity in both children and adults (Laursen et al. 2016; Wolters et al. 2019).

4.5 Gut Microbiota and Obesity Mechanisms

The examination of the implication of the gut microbiota in the pathogenesis of obesity has become a significant focus of research, as documented by Bäckhed et al. (2007). In a complex interplay, it is evident that the gut microbiota might hold the keys to understanding the mechanisms behind obesity. Currently, a considerable amount of research provides evidence for the idea that the gut microbiome exerts a substantial influence on human eating habits (Hanachi et al. 2019). The secretion of gastrointestinal peptides is influenced by commensal bacteria residing in the gastrointestinal tract. Recent findings illustrate the impact of gut microbiome composition and functionality on ghrelinergic signalling through its metabolites, targeting the ghrelin receptor (growth hormone secretagogue receptor). These observations highlight simultaneous changes in circulating ghrelin levels linked to specific alterations in the gut microbiota, which also plays a crucial role in modulating the production of leptin, another hormone associated with appetite. Disruptions in the regulation of these hormones can lead individuals towards increased food consumption and decreased energy expenditure, contributing to the establishment of a pro-obesity environment (Leeuwendaal et al. 2021).

Although not precisely outlined, it is established that the microbiota increases during the initial year of life and undergoes alterations as age advances (Mei et al. 2022), with the bacterial species composition possibly differing based on the presence of obesity (Miranda et al. 2019).

One hypothesised mechanism involves energy harvesting and metabolism, wherein the gut microbiota exerts influence on glucose metabolism and host hormone regulation by generating various metabolites, including but not limited to short-chain fatty acids (SCFA) and bile acids (BA) (Heianza et al. 2018). The fermentation of non-digestible fibres by the microbiota produces SCFAs, serving as an additional energy source. Modification of gut microbiota composition may augment caloric absorption, potentially contributing to weight gain.

Inflammation emerges as another significant factor in this context. Alterations in microbiota can elevate lipopolysaccharide (LPS) levels, inducing chronic low-grade inflammation (Caesar et al. 2012). This metabolic inflammation is closely associated with insulin resistance, a hallmark of complications related to obesity. BA metabolism, crucial for fat digestion, is not exempt from the influence of the microbiota. Changes in gut microbiota can impact BA metabolism, potentially modifying fat absorption and metabolism.

Microbiota-mediated gut barrier dysfunction of the intestinal barrier introduces the concept of a "leaky gut", denoting a state wherein the integrity of the intestinal barrier becomes compromised. A compromised gut barrier permits the passage of substances, including bacteria and their by-products, through the typically selective barrier, allowing them to enter the bloodstream. This breach in the gut barrier has the potential to elicit an inflammatory response and has been correlated with diverse health issues, including complications associated with obesity and implications for overall well-being (Inczefi et al. 2022).

Genetic and environmental factors further complicate the picture. Both genetics and external influences like lifestyle, other than diet, play a role in shaping the gut microbiota. The resulting individual variations may explain why responses to diet and exercise differ across individuals in the context of weight management (Liu et al. 2021).

In conclusion, while the connection between gut microbiota and obesity is a promising avenue of research, it is crucial to recognise the complexity of these interactions. The interplay between genetics, lifestyle, diet, and the gut microbiota forms a dynamic web that we are only beginning to understand. Further research is needed to unravel the intricacies of these mechanisms and explore potential therapeutic interventions for obesity management.

4.6 Definition of Personalised and Sustainable Dietary Patterns Based on Gut Microbiota

Scientific evidence underscores the pivotal role of a "healthy" dietary pattern in shaping a favourable gut microbiota (Lares-Michel et al. 2023). Plant-based, high-fibre diets are associated with anti-inflammatory effects (Trefflich et al. 2020; Sidhu et al. 2023), while Western diets, featuring ultra-processed foods, are linked to dysbiosis (Rinninella et al. 2019). The 2019 definition by the Food and Agriculture Organization of the United Nations and WHO (FAO-WHO 2019) defines sustainable healthy diets as promoting individual health, having a low environmental impact, and being accessible, affordable, safe, equitable, and culturally acceptable. The concept of a "healthy" diet has broadened to encompass sustainability. Achieving a healthy and sustainable diet involves prioritising plant-based, organic, minimally processed foods, as well as incorporating local, seasonal, and fair-trade products (Mazzocchi et al. 2021).

Recent scientific investigations reveal a potential incongruity between a healthy diet and its environmental sustainability (MacDiarmid 2013; Birney et al. 2017; Lares-Michel et al. 2021). Studies indicate that aligning current dietary patterns with recommended guidelines may amplify the environmental impact, leading to a 15% increase in water footprint, 34% rise in energy use, and a 7% surge in food greenhouse gas emissions (Birney et al. 2017). Global dietary guidelines uniformly advocate high consumption of vegetables, fruits, and whole grains, while advising moderation or avoidance of consumption of refined cereals, sugar, saturated fats, and ultra-processed foods (Armet et al. 2022). Guidelines also suggest moderation in the intake of meat, dairy, and eggs (Armet et al. 2022). Notably, diverse environmental impacts exist among food groups, and recent findings emphasise the need for personalised and precise diets (Mekonnen and Hoekstra 2010, 2012; Lonnie et al. 2018; Grasso et al. 2020).

While certain foods associated with lower environmental impact, such as fruits, vegetables, legumes, and whole grains (Willett et al. 2019), are linked to anti-inflammatory bacteria (Trefflich et al. 2020; Sidhu et al. 2023), the environmental implications of other ostensibly healthy foods like nuts and olive oil warrant careful consideration due to their significant water footprint (Blas et al. 2016; Marcelino et al. 2019). For example, while avocado consumption has demonstrated potential health benefits (Thompson et al. 2021), its environmental footprint is pronounced (De la Vega-Rivera and Merino-Pérez 2021). These findings underscore the necessity of a nuanced approach to dietary recommendations that balances health considerations with environmental impact.

Another food group considered essential for the metabolic health of the gut microbiota is dairy products. These foods are included in most dietary guidelines (Armet et al. 2022) as they promote the intestinal microbiota by increasing the beneficial genera such as *Lactobacillus* and *Bifidobacterium*, and reducing potentially pathogenic bacteria such as Bacteroides fragilis (Aslam et al. 2020). However, their role in sustainability remains contentious as dairy is considered one of the food groups with the highest environmental impact (Mekonnen and Hoekstra 2012). In addition to the environmental impact, most dietary guidelines do not consider food costs, indicated as one of the main obstacles to adopting a healthy and sustainable diet (Barosh et al. 2014). Consumption of cheap foods, such as fast food or ultra-processed foods, can alter the gut microbiota, generating dysbiosis (Aguayo-Patrón and Calderón de la Barca 2017). A recent study found that children living in an urban city in the Philippines have a microbiota rich in *Bacteroides* linked to high intakes of lipids, proteins, and free sugars, compared to children living in a rural area whose microbiota is abundant in *Prevotella* (Nakayama et al. 2017). This seems to be related to the fact that urban children have a higher consumption of ultra-processed food (Nakayama et al. 2017). Similar results emerged from the study conducted by De Filippo et al., which compared the intestinal microbiota of European children, who follow a Western diet, with that of children from Burkina Faso, who follow a predominantly rural diet (De Filippo et al. 2010). A study on adults showed that men who consume more than five servings of ultra-processed foods per day present, at the level of the intestinal microbiota, a greater amount of *Granulicatella* (connected with obesity) and *Blautia* (associated with fat accumulation) (Ozato et al. 2019). Although each dietary guideline is country-specific, cultural and social elements are not always considered. Among cultural aspects, the cooking method and recipes specific to each country can alter the composition of the gut microbiota. Cooking increases the digestibility of carbohydrates by gelatinising starch, reducing the amount that reaches the colon, and potentially affecting the fermentation capacity of amylolytic gut bacteria. Cooking methods can also denature antimicrobial compounds naturally present in foods or introduced through agriculture, thus limiting their bioactivity (Carmody et al. 2019). There is therefore a need to stop generalising and to adopt healthy or plant-based diets that are sustainable, but it is important to resort to a contextualised and personalised diet that takes into account the microbiome, the food environment, but also economic, social, and behavioural variables (Lares-Michel et al. 2023). Currently, researchers are increasingly focusing on understanding which food models are certainly healthy and can reduce the impact on the environment. There is no single ecological food model in the world, but in Europe, the Mediterranean diet (MD) and the New Nordic Diet (NND) distinguish themselves as healthy and sustainable dietary patterns (De Cosmi et al. 2022).

The Mediterranean model includes a high consumption of plant foods based on seasonal vegetables and fruits, seeds and cereals (mainly whole grains), nuts, legumes, and extra virgin olive oil as the main source of fat (Davis et al. 2015). It provides for a moderate intake of poultry, eggs, and dairy products (cheese or yogurt), with red wine accompanying meals, and shows a preference for traditional and local products. Fish and seafood replace red meat, which is consumed occasionally and in small quantities (Davis et al. 2015). Numerous studies show that adherence to the MD positively affects the intestinal microbiota, ensuring metabolic health and immune homeostasis (Meslier et al. 2020). A long-term adherence to MD results in an increase in *Roseburia* sp. and *Oscillospira* sp., thereby improving insulin sensitivity in obese patients (Haro et al. 2016; Meslier et al. 2020). Predominantly, *Roseburia* is a butyrate-producing genus that has proven effective in reducing inflammatory effects and is reduced in patients with type 2 diabetes (Haro et al.

2016). On the other hand, higher urinary trimethylamine oxide levels present in individuals with poor adherence to the MD are linked to the risk of cardiovascular diseases (De Filippis et al. 2016). An intervention study on adults identified a decrease in *Haemophilus, Ruminiclostridium, Butyricicoccus, and Eubacterium hallii* in patients consuming the MD compared to the control group (Muralidharan et al. 2021). In addition, a Greek study showed that high adherence to the MD is linked to a decrease in *Escherichia coli* and subsequently to an increase in typical beneficial bifidobacteria (Mitsou et al. 2017). The greater diversity of the intestinal microbiota in subjects following the MD could also be due to the lower presence of processed and animal-based foods compared to Western-type diets (Nagpal et al. 2019). Although numerous studies evaluate the effect of adherence to the MD on the intestinal microbiota in adults, this complex interaction is poorly studied in children and adolescents, especially in those overweight and obese (Haro et al. 2016; De Filippis et al. 2016; Meslier et al. 2020). This may be because studies that evaluated the effect of the MD on adiposity markers and weight maintenance show incomplete results in these age groups compared to those widely established in adults (Trichopoulou et al. 1995; Nagpal et al. 2019).

Finally, it is necessary to emphasise that despite the MD being widely recognised as a healthy and sustainable food model (Hachem et al. 2020), some studies show poor affordability in some socio-economic strata (Rubini et al. 2022) and a high environmental impact on the water footprint (Blas et al. 2016).

Recently, the New Nordic Diet (NND) has been recognised as healthy and also sustainable. It is a typical Northern Europe dietary pattern that involves consuming more calories from vegetable foods and fewer from animal sources, preferring legumes, apples, pears, berries, root vegetables, cabbage, cauliflower, kale, onions, and mushrooms, and whole grains. Fats are mainly derived from nuts, seeds, and canola oil; protein sources from lake and sea fish are preferred; and milk and meat from both wild animals and domestic animals and birds are available. The consumption of processed food, processed meat, and packaged products is not recommended (Mithril et al. 2013; Meltzer et al. 2019). Several studies evaluated the impact of NND on the gut microbiota, indicating positive effects. In a dietary intervention study with the NND, the *Prevotella/Bacteroides* ratio was associated with cholesterol levels in subjects with metabolic syndrome symptoms, suggesting that the microbiota may allow potential control of cholesterol levels (Landberg and Hanhineva 2020). Studies reported that the *Prevotella/Bacteroides* ratio also predicts an individual's ability to lose weight in response to adhering to an NND. Subjects with a higher *Prevotella/Bacteroides* ratio lost more body fat than those with a low ratio following the intake of whole grains and fibre (Hjorth et al. 2018). One of the compounds of microbiotic origin often reported in studies where the NND was evaluated is hippuric acid, which comes from the microbiota-mediated breakdown of dietary phenolic compounds (Lees et al. 2013). One study showed an increase in this acid after a dietary intervention with blueberries, associated with improved glucose homeostasis (de Mello et al. 2017). In addition, hippuric acid has been positively correlated to the species *Faecalibacterium prausnitzii*, which produces butyrate often associated with lower inflammatory markers (van den Munckhof et al.

2018). In addition to hippuric acid, other metabolites derived from the microbiota are related to the Nordic diet, such as furan fatty acids derived from fish. In the recent literature, there are no controlled intervention studies that evaluate the effects of personalised nutritional strategies based on a Nordic diet compared to general recommendations, or whether it is applicable on a large scale (Ordovas et al. 2018; Wang and Hu 2018). Therefore, further studies on this topic are needed.

In light of the above, it can be concluded that through the modulation of the gut microbiome, it is possible not only to regulate the state of health but also to direct towards sustainable choices and preserve biodiversity. The characteristics of a diet for the benefit of the intestinal microbiota coincide with the elements of a healthy food model that guarantees an efficient use of natural resources and a reduced environmental impact. However, further studies, taking into account climatic and soil characteristics, production technologies, and the use of pesticides specific to each region, could provide additional information on their effects on the composition and functions of the gut microbiota to build an increasingly healthy and sustainable food model (Rehner et al. 2023; Biasini et al. 2023).

References

Aguayo-Patrón SV, Calderón de la Barca AM (2017) Old fashioned vs. ultra-processed-based current diets: possible implication in the increased susceptibility to type 1 diabetes and celiac disease in childhood. Food Secur 6

Andela S, Burrows TL, Baur LA et al (2019) Efficacy of very low-energy diet programs for weight loss: a systematic review with meta-analysis of intervention studies in children and adolescents with obesity. Obes Rev 20:871–882. https://doi.org/10.1111/obr.12830

Anderson SE, Cohen P, Naumova EN et al (2007) Adolescent obesity and risk for subsequent major depressive disorder and anxiety disorder: prospective evidence. Psychosom Med 69:740–747. https://doi.org/10.1097/PSY.0b013e31815580b4

Androutsos O, Perperidi M, Georgiou C, Chouliaras G (2021) Lifestyle changes and determinants of children's and adolescents' body weight increase during the first COVID-19 lockdown in Greece: the COV-EAT study. Nutrients 13:1–11. https://doi.org/10.3390/nu13030930

Armet AM, Deehan EC, O'Sullivan AF et al (2022) Rethinking healthy eating in light of the gut microbiome. Cell Host Microbe 30:764–785

Aslam H, Marx W, Rocks T et al (2020) The effects of dairy and dairy derivatives on the gut microbiota: a systematic literature review. Gut Microbes 12(1):1799533. https://doi.org/10.1080/19490976.2020.1799533

Astrup A, Raben A, Geiker N (2015) The role of higher protein diets in weight control and obesity-related comorbidities. Int J Obes 39:721–726. https://doi.org/10.1038/ijo.2014.216

Atlantis E, Barnes EH, Singh MAF (2006) Efficacy of exercise for treating overweight in children and adolescents: a systematic review. Int J Obes 30:1027–1040. https://doi.org/10.1038/sj.ijo.0803286

Ayer J, Charakida M, Deanfield JE, Celermajer DS (2015) Lifetime risk: childhood obesity and cardiovascular risk. Eur Heart J 36:1371–1376. https://doi.org/10.1093/eurheartj/ehv089

Bäckhed F, Manchester JK, Semenkovich CF, Gordon JI (2007) Mechanisms underlying the resistance to diet-induced obesity in germ-free mice. PNAS 104:979–984. https://doi.org/10.1073/pnas.0605374104

Barosh L, Friel S, Engelhardt K, Chan L (2014) The cost of a healthy and sustainable diet – who can afford it? Aust N Z J Public Health 38:7–12. https://doi.org/10.1111/1753-6405.12158

Barrea L, Annunziata G, Bordoni L et al (2020) Nutrigenetics—personalized nutrition in obesity and cardiovascular diseases. Int J Obes Suppl 10:1–13. https://doi.org/10.1038/s41367-020-0014-4

Bassali R, Waller JL, Gower B et al (2010) Utility of waist circumference percentile for risk evaluation in obese children. Int J Pediatr Obes 5:97–101. https://doi.org/10.3109/17477160903111722

Bentham J, Di Cesare M, Bilano V et al (2017) Worldwide trends in body-mass index, underweight, overweight, and obesity from 1975 to 2016: a pooled analysis of 2416 population-based measurement studies in 128·9 million children, adolescents, and adults. Lancet 390:2627–2642. https://doi.org/10.1016/S0140-6736(17)32129-3

Biasini B, Rosi A, Scazzina F, Menozzi D (2023) Predicting the adoption of a sustainable diet in adults: a cross-sectional study in Italy. Nutrients 15. https://doi.org/10.3390/nu15122784

Birney CI, Franklin KF, Davidson FT, Webber ME (2017) An assessment of individual foodprints attributed to diets and food waste in the United States. Environ Res Lett 12. https://doi.org/10.1088/1748-9326/aa8494

Blas A, Garrido A, Willaarts BA (2016) Evaluating the water footprint of the mediterranean and American diets. Water (Switzerland) 8. https://doi.org/10.3390/w8100448

Brambilla P, Pozzobon G, Pietrobelli A (2011) Physical activity as the main therapeutic tool for metabolic syndrome in childhood. Int J Obes 35:16–28. https://doi.org/10.1038/ijo.2010.255

Bray MS, Loos RJF, McCaffery JM et al (2016) NIH working group report – using genomic information to guide weight management: from universal to precision treatment. Obesity 24:14–22. https://doi.org/10.1002/oby.21381

Bryant M, Ashton L, Brown J et al (2014) Systematic review to identify and appraise outcome measures used to evaluate childhood obesity treatment interventions (CoOR): evidence of purpose, application, validity, reliability and sensitivity. Health Technol Assess (Rockv) 18:1–382. https://doi.org/10.3310/hta18510

Burke LE, Wang J, Sevick MA (2011) Self-monitoring in weight loss: a systematic review of the literature. J Am Diet Assoc 111:92–102. https://doi.org/10.1016/j.jada.2010.10.008

Burrows TL, Martin RJ, Collins CE (2010) A systematic review of the validity of dietary assessment methods in children when compared with the method of doubly labeled water. J Am Diet Assoc 110:1501–1510. https://doi.org/10.1016/j.jada.2010.07.008

Burrows T, Janet WM, Collins CE (2011) Long-term changes in food consumption trends in overweight children in the HIKCUPS intervention. J Pediatr Gastroenterol Nutr 53:543–547. https://doi.org/10.1097/MPG.0b013e3182274829

Caesar R, Reigstad CS, Bäckhed HK et al (2012) Gut-derived lipopolysaccharide augments adipose macrophage accumulation but is not essential for impaired glucose or insulin tolerance in mice. Gut 61:1701–1707. https://doi.org/10.1136/gutjnl-2011-301689

Carmody RN, Bisanz JE, Bowen BP et al (2019) Cooking shapes the structure and function of the gut microbiome. Nat Microbiol 4:2052–2063

Chaput JP, Willumsen J, Bull F et al (2020) 2020 WHO guidelines on physical activity and sedentary behaviour for children and adolescents aged 5–17 years: summary of the evidence. Int J Behav Nutr Phys Act 17:1–9. https://doi.org/10.1186/s12966-020-01037-z

Chen S, Dai Y, Ma X et al (2022) Personalized optimal nutrition lifestyle for self obesity management using metaalgorithms. Sci Rep 12:1–12. https://doi.org/10.1038/s41598-022-16260-w

Cho I, Blaser MJ (2012) The human microbiome: at the interface of health and disease. Nat Rev Genet 13:260–270

Cole TJ (1997) Growth monitoring with the British 1990 growth reference. Arch Dis Child 76:47–49. https://doi.org/10.1136/adc.76.1.47

Cronin P, Joyce SA, O'toole PW, O'connor EM (2021) Dietary fibre modulates the gut microbiota. Nutrients 13

Dadhich JP, Gupta A, Rundall P, Bidla N (2019) Interpreting the world health assembly targets on exclusive breastfeeding by 2025: what is expected of each country? World Nutr 10:152–155

David LA, Maurice CF, Carmody RN et al (2014) Diet rapidly and reproducibly alters the human gut microbiome. Nature 505:559–563. https://doi.org/10.1038/nature12820

Davis C, Bryan J, Hodgson J, Murphy K (2015) Definition of the mediterranean diet: a literature review. Nutrients 7:9139–9153

De Cosmi V, Mazzocchi A, Milani GP, Agostoni C (2022) Dietary patterns vs. dietary recommendations. Front Nutr 9:9

De Ferranti SD, Milliren CE, Denhoff ER et al (2015) Providing food to treat adolescents at risk for cardiovascular disease. Obesity 23:2109–2117. https://doi.org/10.1002/oby.21246

De Filippis F, Pellegrini N, Vannini L et al (2016) High-level adherence to a Mediterranean diet beneficially impacts the gut microbiota and associated metabolome. Gut 65:1812–1821. https://doi.org/10.1136/gutjnl-2015-309957

De Filippo C, Cavalieri D, Di Paola M et al (2010) Impact of diet in shaping gut microbiota revealed by a comparative study in children from Europe and rural Africa. PNAS 107(33):14691–14696. https://doi.org/10.1073/pnas.1005963107

De Jesus JM (2011) Expert panel on integrated guidelines for cardiovascular health and risk reduction in children and adolescents: summary report. Pediatrics 128:213–256. https://doi.org/10.1542/peds.2009-2107C

de la Cuesta-Zuluaga J, Kelley ST, Chen Y et al (2019) Age- and sex-dependent patterns of gut microbial diversity in human adults. mSystems 4:4. https://doi.org/10.1128/msystems.00261-19

De la Vega-Rivera A, Merino-Pérez L (2021) Socio-environmental impacts of the avocado boom in the Meseta Purépecha, Michoacán, Mexico. Sustainability (Switzerland) 13. https://doi.org/10.3390/su13137247

de Mello VDF, Lankinen MA, Lindström J et al (2017) Fasting serum hippuric acid is elevated after bilberry (Vaccinium myrtillus) consumption and associates with improvement of fasting glucose levels and insulin secretion in persons at high risk of developing type 2 diabetes. Mol Nutr Food Res 61. https://doi.org/10.1002/mnfr.201700019

De Onis M (2006) WHO child growth standards based on length/height, weight and age. Acta Paediatr 95:76–85. https://doi.org/10.1080/08035320500495548

De Onis M, Onyango AW, Borghi E et al (2007) Development of a WHO growth reference for school-aged children and adolescents. Bull World Health Organ 85:660–667. https://doi.org/10.2471/BLT.07.043497

De Rezende LFM, Lopes MR, Rey-Lopéz JP et al (2014) Sedentary behavior and health outcomes: an overview of systematic reviews. PLoS One 9:e105620. https://doi.org/10.1371/journal.pone.0105620

DeBar LL, Stevens VJ, Perrin N, Philip W, Pearson J, Yarborough BJ, John Dickerson FL (2012) A primary care based, multicomponent lifestyle intervention for overweight adolescent females. Pediatrics 129:e611–e620

Delgado-Noguera M, Tort S, Bonfill X et al (2009) Quality assessment of clinical practice guidelines for the prevention and treatment of childhood overweight and obesity. Eur J Pediatr 168:789–799. https://doi.org/10.1007/s00431-008-0836-5

Dello Russo M, Russo P, Rufián-Henares JÁ et al (2022) The Stance4Health project: evaluating a smart personalised nutrition Service for gut Microbiota Modulation in Normal-and overweight adults and children with obesity, gluten-related disorders or allergy/intolerance to cow's Milk. Food Secur 11:1–10. https://doi.org/10.3390/foods11101480

Derks IPM, Kocevska D, Jaddoe VWV et al (2017) Longitudinal associations of sleep duration in infancy and early childhood with body composition and Cardiometabolic health at the age of 6 years: the generation R study. Child Obes 13:400–408. https://doi.org/10.1089/chi.2016.0341

Di Bonito P, Pacifico L, Chiesa C et al (2017) Impaired fasting glucose and impaired glucose tolerance in children and adolescents with overweight/obesity. J Endocrinol Investig 40:409–416. https://doi.org/10.1007/s40618-016-0576-8

Dominika Ś, Arjan N, Karyn RP, Henryk K (2011) The study on the impact of glycated pea proteins on human intestinal bacteria. Int J Food Microbiol 145:267–272. https://doi.org/10.1016/j.ijfoodmicro.2011.01.002

DSAM (n.d.) Opsporing og behandling af overvægt hos førskolebørn DSAM og Sundhedsstyrelsen. http://www.sst.dk/publ/Publ2006/CFF/Overvaegt/Overvaegtigeboern.pdf

Duncanson K, Shrewsbury V, Collins C (2017) Interim report on the effectiveness of dietary interventions for children and adolescents with overweight and obesity for the World Health Organization

Duncanson K, Shrewsbury V, Burrows T et al (2021) Impact of weight management nutrition interventions on dietary outcomes in children and adolescents with overweight or obesity: a systematic review with meta-analysis. J Hum Nutr Diet 34:147–177. https://doi.org/10.1111/jhn.12831

Ebbeling CB, Feldman HA, Chomitz VR et al (2012) A randomized trial of sugar-sweetened beverages and adolescent body weight. N Engl J Med 367:1407–1416. https://doi.org/10.1056/nejmoa1203388

Ells LJ, Rees K, Brown T et al (2018) Interventions for treating children and adolescents with overweight and obesity: an overview of Cochrane reviews. Int J Obes 42:1823–1833. https://doi.org/10.1038/s41366-018-0230-y

Epstein LH, Paluch RA, Beecher MD, Roemmich JN (2008) Increasing healthy eating vs. reducing high energy-dense foods to treat pediatric obesity. Obesity 16:318–326. https://doi.org/10.1038/oby.2007.61

Epstein LH, Kilanowski C, Paluch RA et al (2015) Reducing variety enhances effectiveness of family-based treatment for pediatric obesity. Eat Behav 17:140–143. https://doi.org/10.1016/j.eatbeh.2015.02.001

Esfahani A, Wong JMW, Mirrahimi A et al (2011) The application of the glycemic index and glycemic load in weight loss: a review of the clinical evidence. IUBMB Life 63:7–13. https://doi.org/10.1002/iub.418

FAO and WHO (2019) Sustainable healthy diets – guiding principles. Rome. https://www.researchgate.net/publication/337136307

Fatima Y, Doi SAR, Mamun AA (2015) Longitudinal impact of sleep on overweight and obesity in children and adolescents: a systematic review and bias-adjusted meta-analysis. Obes Rev 16: 137–149. https://doi.org/10.1111/obr.12245

Faulkner MS, Michaliszyn SF, Hepworth JT, Wheeler MD (2014) Personalized exercise for adolescents with diabetes or obesity. Biol Res Nurs 16:46–54. https://doi.org/10.1177/1099800413500064

Finne E, Reinehr T, Schaefer A et al (2013) Changes in self-reported and parent-reported health-related quality of life in overweight children and adolescents participating in an outpatient training: findings from a 12-month follow-up study. Health Qual Life Outcomes 11:1. https://doi.org/10.1186/1477-7525-11-1

Food and Agriculture Organization of the United Nations (2022) Food-based dietary guidelines. https://www.fao.org/nutrition/education/food-dietary-guidelines/regions/en/

Friedemann C, Heneghan C, Mahtani K et al (2012) Cardiovascular disease risk in healthy children and its association with body mass index: systematic review and meta-analysis. BMJ (Online) 345:1–16. https://doi.org/10.1136/bmj.e4759

Gao Z, Chen S (2014) Are field-based exergames useful in preventing childhood obesity? A systematic review. Obes Rev 15:676–691. https://doi.org/10.1111/obr.12164

Ghanvatkar S, Kankanhalli A, Rajan V (2019) User models for personalized physical activity interventions: scoping review. JMIR Mhealth Uhealth 7:e11098. https://doi.org/10.2196/11098

Gibson AA, Sainsbury A (2017) Strategies to improve adherence to dietary weight loss interventions in research and real-world settings. Behav Sci 7. https://doi.org/10.3390/bs7030044

Gill N, Gjelsvik A, Mercurio LY, Amanullah S (2021) Childhood obesity is associated with poor academic skills and coping mechanisms. J Pediatr 228:278–284. https://doi.org/10.1016/j.jpeds.2020.09.004

Global Health Observatory (n.d.) Global health observatory. Geneva: World Health or https://www.who.int/data/gho/data/themes/mortality-and-global-health-estimates

Goss AM, Dowla S, Pendergrass M et al (2020) Effects of a carbohydrate-restricted diet on hepatic lipid content in adolescents with non-alcoholic fatty liver disease: a pilot, randomized trial. Pediatr Obes 15:1–10. https://doi.org/10.1111/ijpo.12630

Gow ML, Garnett SP, Baur LA, Lister NB (2016) The effectiveness of different diet strategies to reduce type 2 diabetes risk in youth. Nutrients 8:1–13. https://doi.org/10.3390/nu8080486

Gow ML, Baur LA, Johnson NA et al (2017) Reversal of type 2 diabetes in youth who adhere to a very-low-energy diet: a pilot study. Diabetologia 60:406–415. https://doi.org/10.1007/s00125-016-4163-5

Grasso AC, Olthof MR, van Dooren C et al (2020) Effect of food-related behavioral activation therapy on food intake and the environmental impact of the diet: results from the MooDFOOD prevention trial. Eur J Nutr 59:2579–2591. https://doi.org/10.1007/s00394-019-02106-1

Hachem F, Vanham D, Moreno LA (2020) Territorial and sustainable healthy diets. Food Nutr Bull 41:87S–103S. https://doi.org/10.1177/0379572120976253

Hanachi M, Manichanh C, Schoenenberger A et al (2019) Altered host-gut microbes symbiosis in severely malnourished anorexia nervosa (AN) patients undergoing enteral nutrition: an explicative factor of functional intestinal disorders? Clin Nutr 38:2304–2310. https://doi.org/10.1016/j.clnu.2018.10.004

Haro C, Montes-Borrego M, Rangel-Zúñiga OA et al (2016) Two healthy diets modulate gut microbial community improving insulin sensitivity in a human obese population. J Clin Endocrinol Metab 101:233–242. https://doi.org/10.1210/jc.2015-3351

Hassapidou M, Duncanson K, Shrewsbury V et al (2023) EASO and EFAD position statement on medical nutrition therapy for the Management of Overweight and Obesity in children and adolescents. Obes Facts 16:29–52. https://doi.org/10.1159/000527540

Healey G, Murphy R, Butts C et al (2018) Habitual dietary fibre intake influences gut microbiota response to an inulin-type fructan prebiotic: a randomised, double-blind, placebo-controlled, cross-over, human intervention study. Br J Nutr 119:176–189. https://doi.org/10.1017/S0007114517003440

Heianza Y, Sun D, Smith SR et al (2018) Changes in gut microbiota- related metabolites and long-term successful weight loss in response to weight-loss diets: the POUNDS lost trial. Diabetes Care 41:413–419. https://doi.org/10.2337/dc17-2108

Hernández M, Castellet J, Narvaiza JL, Rincón JM et al (1988) Growth curves and tables Madrid: Instituto de Investigación sobre Crecimiento y Desarrollo

Hjorth MF, Roager HM, Larsen TM et al (2018) Pre-treatment microbial Prevotella-to-Bacteroides ratio determines body fat loss success during a 6-month randomized controlled diet intervention. Int J Obes 42:580–583. https://doi.org/10.1038/ijo.2017.220

Holzapfel C, Drabsch T (2019) A scientific perspective of personalised gene-based dietary recommendations for weight management. Nutrients 11:1–14. https://doi.org/10.3390/nu11030617

Inczefi O, Bacsur P, Resál T et al (2022) The influence of nutrition on intestinal permeability and the microbiome in health and disease. Front Nutr 9

Institute of Medicine (U.S.) (2011) Committee on standards for developing trustworthy clinical practice guidelines. Clinical practice guidelines we can trust. The National Academies Press (U.S.), Washington

Janssen I (2007) Physical activity guidelines for children and youth. Appl Physiol Nutr Metab 32: S109–S121

Janssen I, LeBlanc AG (2010) Systematic review of the health benefits of physical activity and fitness in school-aged children and youth. Int J Behav Nutr Phys Act 7:40. https://doi.org/10.1186/1479-5868-7-40

Jebeile H, Kelly AS, O'Malley G, Baur LA (2022) Obesity in children and adolescents: epidemiology, causes, assessment, and management. Lancet Diabetes Endocrinol 10:351–365. https://doi.org/10.1016/S2213-8587(22)00047-X

Johansen A, Holm JC, Pearson S et al (2015) Danish clinical guidelines for examination and treatment of overweight and obese children and adolescents in a pediatric setting. Dan Med J 62

Kelley GA, Kelley KS, Pate RR (2015) Exercise and BMI in overweight and obese children and adolescents: a systematic review and trial sequential meta-analysis. Biomed Res Int 2015(1): 704539. https://doi.org/10.1155/2015/704539

Khan NA, Baym CL, Monti JM et al (2015) Central adiposity is negatively associated with hippocampal-dependent relational memory among overweight and obese children. J Pediatr 166:302–308.e1. https://doi.org/10.1016/j.jpeds.2014.10.008

Khan NA, Cannavale C, Iwinski S et al (2020) Visceral adiposity and diet quality are differentially associated with cognitive abilities and early academic skills among preschool-age children. Front Pediatr 7:1–10. https://doi.org/10.3389/fped.2019.00548

Kim J, Kim YM, Jang HB et al (2019) Evidence-based nutritional intervention protocol for Korean moderate-severe obese children and adolescents. Clin Nutr Res 8:184. https://doi.org/10.7762/cnr.2019.8.3.184

Kirk S, Brehm B, Saelens BE et al (2012) Role of carbohydrate modification in weight management among obese children: a randomized clinical trial. J Pediatr 161:320. https://doi.org/10.1016/j.jpeds.2012.01.041

Kirschenbaum DS, Gierut K (2013) Treatment of childhood and adolescent obesity: an integrative review of recent recommendations from five expert groups. J Consult Clin Psychol 81:347–360. https://doi.org/10.1037/a0030497

Koliaki C, Spinos T, Spinou M et al (2018) Defining the optimal dietary approach for safe, effective and sustainable weight loss in overweight and obese adults. Healthcare (Switzerland) 6. https://doi.org/10.3390/healthcare6030073

Kolotourou M, Radley D, Chadwick P et al (2013) Is BMI alone a sufficient outcome to evaluate interventions for child obesity? Child Obes 9:350–356. https://doi.org/10.1089/chi.2013.0019

Korem T, Zeevi D, Zmora N et al (2017) Bread affects clinical parameters and induces gut microbiome-associated personal glycemic responses. Cell Metab 25:1243–1253.e5. https://doi.org/10.1016/j.cmet.2017.05.002

Koulougliotі C, Cole R, McQuillan B et al (2013) Feasibility of an individualized, home-based obesity prevention program for preschool-age children. Child Health Care 42:134–152. https://doi.org/10.1080/02739615.2013.766099

Landberg R, Hanhineva K (2020) Biomarkers of a healthy Nordic diet – from dietary exposure biomarkers to microbiota signatures in the metabolome. Nutrients 12

Lares-Michel M, Housni FE, Aguilera Cervantes VG et al (2021) Eat well to fight obesity… and save water: the water footprint of different diets and caloric intake and its relationship with adiposity. Front Nutr 8. https://doi.org/10.3389/fnut.2021.694775

Lares-Michel M, Reyes-Castillo Z, Housni FE (2023) Towards the characterisation of sustainable diet's gut microbiota composition and functions: a narrative review. Gut Microbiome 4:e15. https://doi.org/10.1017/gmb.2023.13

Larsen LM, Hertel NT, Mølgaard C et al (2015) Early intervention for childhood overweight: a randomized trial in general practice. Scand J Prim Health Care 33:184–190. https://doi.org/10.3109/02813432.2015.1067511

Larsen KT, Huang T, Ried-Larsen M et al (2016) A multi-component day-camp weight-loss program is effective in reducing BMI in children after one year: a randomized controlled trial. PLoS One 11:1–16. https://doi.org/10.1371/journal.pone.0157182

Laursen MF, Andersen LBB, Michaelsen KF et al (2016) Infant gut microbiota development is driven by transition to family foods independent of maternal obesity. mSphere 1. https://doi.org/10.1128/msphere.00069-15

Le Chatelier E, Nielsen T, Qin J et al (2013) Richness of human gut microbiome correlates with metabolic markers. Nature 500:541–546. https://doi.org/10.1038/nature12506

LeBlanc AG, Spence JC, Carson V et al (2012) Systematic review of sedentary behaviour and health indicators in the early years (aged 0-4 years). Appl Physiol Nutr Metab 37:753–772. https://doi.org/10.1139/H2012-063

Lee YS, Biddle S, Chan MF et al (2016) Health promotion board–ministry of health clinical practice guidelines: obesity. Singapore Med J 57:292–300. https://doi.org/10.11622/smedj.2016103

Leeming ER, Johnson AJ, Spector TD, Roy CIL (2019) Effect of diet on the gut microbiota: rethinking intervention duration. Nutrients 11

Lees HJ, Swann JR, Wilson ID et al (2013) Hippurate: the natural history of a mammalian-microbial cometabolite. J Proteome Res 12:1527–1546. https://doi.org/10.1021/pr300900b

Leeuwendaal NK, Cryan JF, Schellekens H (2021) Gut peptides and the microbiome: focus on ghrelin. Curr Opin Endocrinol Diabetes Obes 28:243–252

Liu BN, Liu XT, Liang ZH, Wang JH (2021) Gut microbiota in obesity. World J Gastroenterol 27: 3837–3850

Lonnie M, Hooker E, Brunstrom JM et al (2018) Protein for life: review of optimal protein intake, sustainable dietary sources and the effect on appetite in ageing adults. Nutrients 10

Looney SM, Raynor HA (2014) Examining the effect of three low-intensity pediatric obesity interventions: a pilot randomized controlled trial. Clin Pediatr (Phila) 53:1367–1374. https://doi.org/10.1177/0009922814541803

Losno EA, Sieferle K, Perez-Cueto FJA, Ritz C (2021) Vegan diet and the gut microbiota composition in healthy adults. Nutrients 13. https://doi.org/10.3390/nu13072402

Lurbe E, Agabiti-Rosei E, Cruickshank JK et al (2016) 2016 European Society of hypertension guidelines for the management of high blood pressure in children and adolescents

MacDiarmid JI (2013) Is a healthy diet an environmentally sustainable diet? In: Proceedings of the nutrition society, pp 13–20

Macknin M, Kong T, Weier A et al (2015) Plant-based, no-added-fat or American heart association diets: impact on cardiovascular risk in obese children with hypercholesterolemia and their parents. J Pediatr 166:953–959.e3. https://doi.org/10.1016/j.jpeds.2014.12.058

Marcelino G, Hiane PA et al (2019) Effects of olive oil and its minor components on cardiovascular diseases, inflammation, and gut microbiota. Nutrients 11

Martin A, Booth JN, McGeown S et al (2017) Longitudinal associations between childhood obesity and academic achievement: systematic review with focus group data. Curr Obes Rep 6:297–313. https://doi.org/10.1007/s13679-017-0272-9

Mazzocchi A, De Cosmi V, Scaglioni S, Agostoni C (2021) Towards a more sustainable nutrition: complementary feeding and early taste experiences as a basis for future food choices. Nutrients 13. https://doi.org/10.3390/nu13082695

McGovern L, Johnson JN, Paulo R et al (2008) Treatment of pediatric obesity: a systematic review and meta-analysis of randomized trials. J Clin Endocrinol Metab 93:4600–4605. https://doi.org/10.1210/jc.2006-2409

McNarry MA, Mackintosh KA (2016) Investigating the relative exercise intensity of Exergames in Prepubertal children. Games Health J 5:135–140. https://doi.org/10.1089/g4h.2015.0094

Mead E, Brown T, Rees K, Azevedo LB, Whittaker V, Jones D et al (2017) Diet, physical activity and behavioural interventions for the treatment of overweight or obese children from the age of 6 to 11 years. Cochrane Database Syst Rev 6

Mei H, Li N, Zhang Y et al (2022) Gut microbiota diversity and overweight/obesity in infancy: results from a nested case-control study. Curr Med Sci 42:210–216. https://doi.org/10.1007/s11596-021-2476-1

Mekonnen MM, Hoekstra AY (2010) Value of water research report series no. 47 value of water

Mekonnen MM, Hoekstra AY (2012) A global assessment of the water footprint of farm animal products. Ecosystems 15:401–415. https://doi.org/10.1007/s10021-011-9517-8

Meltzer HM, Brantsæter AL, Trolle E et al (2019) Environmental sustainability perspectives of the nordic diet. Nutrients 11. https://doi.org/10.3390/nu11092248

Meslier V, Laiola M, Roager HM et al (2020) Mediterranean diet intervention in overweight and obese subjects lowers plasma cholesterol and causes changes in the gut microbiome and metabolome independently of energy intake. Gut 69:1258–1268. https://doi.org/10.1136/gutjnl-2019-320438

Michigan Quality Improvement Consortium (MQIC) (2018) Treatment of childhood overweight and obesity. http://www.mqic.org/pdf/2018_mqic_treatment_of_childhood_overweight_and_obesity_cpg.pdf

Miranda VPN, Dos Santos Amorim PR, Bastos RR et al (2019) Abundance of gut microbiota, concentration of short-chain fatty acids, and inflammatory markers associated with elevated body fat, overweight, and obesity in female adolescents. Mediat Inflamm 2019:1. https://doi.org/10.1155/2019/7346863

Mirza NM, Palmer MG, Sinclair KB et al (2013) Effects of a low glycemic load or a low-fat dietary intervention on body weight in obese Hispanic American children and adolescents: a randomized controlled trial. Am J Clin Nutr 97:276–285. https://doi.org/10.3945/ajcn.112.042630

Mithril C, Dragsted LO, Meyer C et al (2013) Dietary composition and nutrient content of the new Nordic diet. Public Health Nutr 16:777–785. https://doi.org/10.1017/S1368980012004521

Mitsou EK, Kakali A, Antonopoulou S et al (2017) Adherence to the Mediterranean diet is associated with the gut microbiota pattern and gastrointestinal characteristics in an adult population. Br J Nutr 117:1645–1655. https://doi.org/10.1017/S0007114517001593

Monasta L, Lobstein T, Cole TJ et al (2011) Defining overweight and obesity in pre-school children: IOTF reference or WHO standard? Obes Rev 12:295–300. https://doi.org/10.1111/j.1467-789X.2010.00748.x

Motevalli M, Drenowatz C, Tanous DR et al (2021) Management of childhood obesity – time to shift from generalized to personalized intervention strategies. Nutrients 13. https://doi.org/10.3390/nu13041200

Muralidharan J, Moreno-Indias I, Bulló M et al (2021) Effect on gut microbiota of a 1-y lifestyle intervention with Mediterranean diet compared with energy-reduced Mediterranean diet and physical activity promotion: PREDIMED-plus study. Am J Clin Nutr 114:1148–1158. https://doi.org/10.1093/ajcn/nqab150

Nagpal R, Shively CA, Register TC et al (2019) Gut microbiome-mediterranean diet interactions in improving host health. F1000Res 8:699. https://doi.org/10.12688/F1000RESEARCH.18992.1

Nakayama J, Yamamoto A, Palermo-Conde LA et al (2017) Impact of westernized diet on gut microbiota in children on Leyte Island. Front Microbiol 8. https://doi.org/10.3389/fmicb.2017.00197

National High Blood Pressure Education Program Working Group on High Blood Pressure in Children and Adolescents (2004) The fourth report on the diagnosis, evaluation, and treatment of high blood pressure in children and adolescents. Pediatrics 114:555–576

Noce A, Marrone G, Di Daniele F et al (2019) Impact of gut microbiota composition on onset and progression of chronic non-communicable diseases. Nutrients 11. https://doi.org/10.3390/nu11051073

O'Malley GC, Shultz SP, Thivel D, Tsiros MD (2021) Neuromusculoskeletal health in pediatric obesity: incorporating evidence into clinical examination. Curr Obes Rep 10:467–477. https://doi.org/10.1007/s13679-021-00463-9

Ordovas JM, Ferguson LR, Tai ES, Mathers JC (2018) Personalised nutrition and health. BMJ (Online) 361:bmj.k2173. https://doi.org/10.1136/bmj.k2173

Oude Luttikhuis H, Baur L, Jansen H, Shrewsbury VA, O'Malley C, Stolk RPSC (2009) Interventions for treating obesity in children. Cochrane Database Syst Rev

Ozato N, Saito S, Yamaguchi T et al (2019) Blautia genus associated with visceral fat accumulation in adults 20–76 years of age. NPJ Biofilms Microbiomes 5:28. https://doi.org/10.1038/s41522-019-0101-x

Parekh PJ, Arusi E, Vinik AI, Johnson DA (2014) The role and influence of gut microbiota in pathogenesis and management of obesity and metabolic syndrome. Front Endocrinol (Lausanne) 5

Pearson N, Biddle SJH (2011) Sedentary behavior and dietary intake in children, adolescents, and adults: a systematic review. Am J Prev Med 41:178–188. https://doi.org/10.1016/j.amepre.2011.05.002

Peterson AL, McBride PE (2012) A review of guidelines for dyslipidemia in children and adolescents. Wis Med J 111:274–281

Pfeifflé S, Pellegrino F, Kruseman M et al (2019) Current recommendations for nutritional management of overweight and obesity in children and adolescents: a structured framework. Nutrients 11. https://doi.org/10.3390/nu11020362

Pietrobelli A, Pecoraro L, Ferruzzi A et al (2020) Effects of COVID-19 lockdown on lifestyle behaviors in children with obesity living in Verona, Italy: a longitudinal study. Obesity 28: 1382–1385. https://doi.org/10.1002/oby.22861

Rajjo T, Mohammed K, Alsawas M et al (2017) Treatment of pediatric obesity: an umbrella systematic review. J Clin Endocrinol Metab 102:763–775. https://doi.org/10.1210/jc.2016-2574

Rehner J, Schmartz GP, Kramer T et al (2023) The effect of a planetary health diet on the human gut microbiome: a descriptive analysis. Nutrients 15. https://doi.org/10.3390/nu15081924

Richardson L, Paulis WD, van Middelkoop M, Koes BW (2013) An overview of national clinical guidelines for the management of childhood obesity in primary care. Prev Med (Baltim) 57: 448–455. https://doi.org/10.1016/j.ypmed.2013.08.010

Rinninella E, Raoul P, Cintoni M et al (2019) What is the healthy gut microbiota composition? A changing ecosystem across age, environment, diet, and diseases. Microorganisms 7. https://doi.org/10.3390/microorganisms7010014

Rothschild D, Weissbrod O, Barkan E et al (2018) Environment dominates over host genetics in shaping human gut microbiota. Nature 555:210–215. https://doi.org/10.1038/nature25973

Rubini A, Vilaplana-Prieto C, Flor-Alemany M et al (2022) Assessment of the cost of the Mediterranean diet in a low-income region: adherence and relationship with available incomes. BMC Public Health 22. https://doi.org/10.1186/s12889-021-12433-w

Saha AK, Sarkar N, Chatterjee T (2011) Health consequences of childhood obesity. Indian J Pediatr 78:1349–1355. https://doi.org/10.1007/s12098-011-0489-7

Sarfraz MH, Shahid A, Asghar S et al (2022) Personalized nutrition, microbiota, and metabolism: a triad for eudaimonia. Front Mol Biosci 9

Scottish Intercollegiate Guidelines Network (2010) Management of obesity – a national clinical guideline (SIGN Guideline No 115). Edinburgh (U.K.)

Sidhu SRK, Kok CW, Kunasegaran T, Ramadas A (2023) Effect of plant-based diets on gut microbiota: a systematic review of interventional studies. Nutrients 15

Silva DAS, Lang JJ, Barnes JD et al (2018) Cardiorespiratory fitness in children: evidence for criterion-referenced cut-points. PLoS One 13:15–17. https://doi.org/10.1371/journal.pone.0201048

Singh RK, Chang HW, Yan D et al (2017) Influence of diet on the gut microbiome and implications for human health. J Transl Med 15

Singhal V, Sella AC, Malhotra S (2021) Pharmacotherapy in pediatric obesity: current evidence and landscape. Curr Opin Endocrinol Diabetes Obes 28:55–63

Small L, Professor A, Melnyk B et al (2015) The preliminary effects of a primary care–based randomized treatment trial with overweight and obese young children and their parents. J Pediatr Health Care 28:602–496. https://doi.org/10.1016/j.pedhc.2013.01.003

Spanish Ministry for Health (2009) Working group of the guideline for the prevention and treatment of childhood and juvenile obesity. Clinical practice guideline for the prevention and treatment of childhood and juvenile obesity; Quality plan for the Spanish national healthcare system of the Spanish ministry for health and social policy. Available online at: https://portal.guiasalud.es/wp-content/uploads/2019/01/GPC_452_obes_infantojuv_AATRM_compl_en.pdf. Accessed Feb 2024

Strong WB, Malina RM, Blimkie CJR et al (2005) Evidence based physical activity for school-age youth. J Pediatr 146:732–737. https://doi.org/10.1016/j.jpeds.2005.01.055

Styne DM, Arslanian SA, Connor EL et al (2017) Pediatric obesity-assessment, treatment, and prevention: an endocrine society clinical practice guideline. J Clin Endocrinol Metab 102:709–757. https://doi.org/10.1210/jc.2016-2573

Te Velde SJ, Van Nassau F, Uijtdewilligen L et al (2012) Energy balance-related behaviours associated with overweight and obesity in preschool children: a systematic review of prospective studies. Obes Rev 13:56–74. https://doi.org/10.1111/j.1467-789X.2011.00960.x

Thompson SV, Bailey MA, Taylor AM et al (2021) Avocado consumption alters gastrointestinal bacteria abundance and microbial metabolite concentrations among adults with overweight or obesity: a randomized controlled trial. J Nutr 151:753–762. https://doi.org/10.1093/jn/nxaa219

Tinggaard J, Aksglaede L, Sørensen K et al (2014) The 2014 Danish references from birth to 20 years for height, weight and body mass index. Acta Paediatr 103:214–224. https://doi.org/10.1111/apa.12468

Tomayko EJ, Prince RJ, Cronin KA, Adams AK (2016) The healthy children, Strong families intervention promotes improvements in nutrition, activity and body weight in American Indian families with young children. Public Health Nutr 19:2850–2859. https://doi.org/10.1017/S1368980016001014

Tomova A, Bukovsky I, Rembert E et al (2019) The effects of vegetarian and vegan diets on gut microbiota. Front Nutr 6

Trefflich I, Jabakhanji A, Menzel J et al (2020) Is a vegan or a vegetarian diet associated with the microbiota composition in the gut? Results of a new cross-sectional study and systematic review. Crit Rev Food Sci Nutr 60:2990–3004

Trichopoulou A, Kouris-Blazos A, Wahlqvist ML, Gnardellis C, Lagiou P, Polychronopoulos E et al (1995) Diet and overall survival in elderly people. BMJ 311(7018):1457–1460

Valerio G, Maffeis C, Saggese G et al (2018) Diagnosis, treatment and prevention of pediatric obesity: consensus position statement of the Italian Society for Pediatric Endocrinology and Diabetology and the Italian Society of Pediatrics. Ital J Pediatr 44:1–21. https://doi.org/10.1186/s13052-018-0525-6

van den Munckhof ICL, Kurilshikov A, ter Horst R et al (2018) Role of gut microbiota in chronic low-grade inflammation as potential driver for atherosclerotic cardiovascular disease: a systematic review of human studies. Obes Rev 19:1719–1734

Vandewater EA, Park SE, Hébert ET, Cummings HM (2015) Time with friends and physical activity as mechanisms linking obesity and television viewing among youth. Int J Behav Nutr Phys Act 12:S6. https://doi.org/10.1186/1479-5868-12-S1-S6

Vanegas SM, Meydani M, Barnett JB et al (2017) Substituting whole grains for refined grains in a 6-wk randomized trial has a modest effect on gut microbiota and immune and inflammatory markers of healthy adults. Am J Clin Nutr 105:635–650. https://doi.org/10.3945/ajcn.116.146928

Verrotti A, Agostinelli S, D'Egidio C et al (2013) Impact of a weight loss program on migraine in obese adolescents. Eur J Neurol 20:394–397. https://doi.org/10.1111/j.1468-1331.2012.03771.x

Vlassopoulos A, Govers E, Mulrooney H et al (2021) Dietetic management of obesity in Europe: gaps in current practice. Eur J Clin Nutr 75:1155–1158. https://doi.org/10.1038/s41430-020-00820-2

Wang DD, Hu FB (2018) Precision nutrition for prevention and management of type 2 diabetes. Lancet Diabetes Endocrinol 6:416–426

Weihe P, Spielmann J, Kielstein H et al (2020) Childhood obesity and cancer risk in adulthood. Curr Obes Rep 9:204–212. https://doi.org/10.1007/s13679-020-00387-w

Willett W, Rockström J, Loken B et al (2019) Food in the Anthropocene: the EAT–lancet commission on healthy diets from sustainable food systems. Lancet 393:447–492

Willi SM, Martin K, Datko FM, Brant BP (2004) Treatment of type 2 diabetes in childhood using a very-low-calorie diet. Diabetes Care 27:348–353. https://doi.org/10.2337/diacare.27.2.348

Wolters M, Ahrens J, Romaní-Pérez M et al (2019) Dietary fat, the gut microbiota, and metabolic health – a systematic review conducted within the MyNewGut project. Clin Nutr 38:2504–2520

Woolf SH, Grol R, Hutchinson A et al (1999) Clinical guidelines potential benefits, limitations, and harms of clinical guidelines. Br Med J 318:527

World Health Organization (2022) Regional Office for Europe WHO European Regional Obesity: Report 2022

World Health Organization. Childhood Obesity Surveillance Initiative (2021) World Health Organization. Childhood Obesity Survey. https://www.euro.who.int/en/health-topics/disease-prevention/nutrition/activities/who-europeanchildhood-obesity-surveillance-initiative-cosi

Wu GD, Chen J, Hoffmann C et al (2011) Linking long-term dietary patterns with gut microbial enterotypes. Science 334:105–108. https://doi.org/10.1126/science.1208344

Wühl E (2019) Hypertension in childhood obesity. Acta Paediatr 108:37–43. https://doi.org/10.1111/apa.14551

Zeevi D, Korem T, Zmora N et al (2015) Personalized nutrition by prediction of glycemic responses. Cell 163:1079–1094. https://doi.org/10.1016/j.cell.2015.11.001

Zeinalian R, Ahmadikhatir S, Esfahani EN et al (2022) The roles of personalized nutrition in obesity and diabetes management: a review. J Diabetes Metab Disord 21:1119–1127. https://doi.org/10.1007/s40200-022-01016-0

Nutrition, Microbiome, and Allergy

5

Dafni Moriki, Kostas N. Priftis, and Konstantinos Douros

Abstract

A growing body of evidence suggests that the establishment of a balanced gut microbiome is crucial for many aspects of human health. This is because gut microbes are involved in key processes such as host metabolism and immune system regulation and therefore help to maintain homeostasis in the human body. The gut microbiome and the immune system evolve in parallel, and their interactions are established during infancy and early childhood. Altered gut colonization patterns and reduced gut diversity during early life have been associated with the development of several diseases, including allergies, later in life. Undoubtedly, nutrition is a key regulator of the gut microbiome and plays a vital role in the development of such diseases. In this chapter, we will discuss how the gut microbiome interacts with the immune system in order to achieve immune tolerance. We will also describe the factors that influence the gut colonization process during early life and how disruption of this process can lead to the development of allergic diseases. Finally, we will focus on the role of nutrition during pregnancy, infancy, and early childhood in the development of the immune system. In particular, we will focus on the role of certain nutrients and food components, such as human milk oligosaccharides (HMOs), short-chain fatty acids (SCFAs), and omega-3 polyunsaturated fatty acids (PUFAs).

Keywords

Gut · Microbiome · Diet · Nutrition · Allergy · HMOs · SCFAs · Omega-3 PUFAs

D. Moriki (✉) · K. N. Priftis · K. Douros
Pediatric Allergy and Respiratory Unit, 3rd Department of Pediatrics, "Attikon" University Hospital, School of Medicine, National and Kapodistrian University of Athens, Athens, Greece

M. P. Francino, J. Á. Rufián Henares (eds.), *The Microbiome and Personalized Nutrition*, The Microbiomes of Humans, Animals, Plants, and the Environment 6,
https://doi.org/10.1007/978-3-031-93902-0_5

5.1 Allergic Diseases and Underlying Immune Mechanisms

Allergy is an inappropriate immune response to specific substances, called allergens, that are usually harmless to most people. Such substances are mainly food proteins and inhalant allergens such as pollen, mold, animal dander, and house dust mites. More rarely, drugs and venoms can also cause allergic immune responses and are therefore considered allergens. The human body is exposed to a large amount of allergens through the epithelial surfaces of the body, such as the skin, the upper and lower airways, and the gastrointestinal tract. The epithelium of such surfaces consists of a thin layer of cells that acts as a physical barrier between the environment and the human body and constitutes the first line of defense (Larsen et al. 2020). Epithelial cells are equipped with a variety of pattern recognition receptors (PRRs) that recognize molecules found on pathogens (pathogen-associated molecular patterns [PAMPs]) or molecules released from damaged cells (damage-associated molecular patterns [DAMPs]) and are able to communicate with the immune system to enable coordinated responses to maintain host homeostasis and mount host defense (Li and Wu 2021). Mucosal epithelial barriers, such as the intestinal and airway epithelial barriers, are also composed of mucus, microbiome, surface liquids, and junctional complexes that reinforce barrier integrity (Akdis 2021). Dysfunction of these highly organized barriers allows penetration of allergens and leads to the release of type 2 cytokines that drive allergic inflammation and promote the development of allergic diseases (Pulendran and Artis 2012).

Allergic immune reactions can cause a variety of symptoms including hives, angioedema, coughing, shortness of breath, sneezing, hypotension, and vomiting. The manifestations of allergic diseases may appear at any age but most often occur in infancy and early childhood due to the immature immune system. The most common allergic diseases include asthma, allergic rhinitis, atopic dermatitis, food allergy, and anaphylaxis. Atopic dermatitis and food allergy are prevalent in infancy, while asthma and allergic rhinitis usually develop in school-age children. In fact, allergic diseases often follow an ordered developmental progression (from atopic dermatitis through food allergy to inhalant allergy) referred to as the "atopic march" (Spergel and Paller 2003). The underlying immune mechanism of most allergic diseases is mediated by immunoglobulin E (IgE). Other immune mechanisms, independent of IgE, have also been described but are less common. In the case of IgE-mediated immune responses, the immunological process presents in two phases: the sensitization phase and the effector phase. The sensitization phase is initiated when an allergen binds to high-affinity receptors for IgE (FcεRI) on antigen-presenting cells (APCs), such as dendritic cells (DCs) and macrophages (Novak et al. 2003). APCs then process the allergen into small peptides that are presented for recognition by naïve T lymphocytes (T cells) via the major histocompatibility complex (MHC) class II molecules. This results in the differentiation of T cells into T helper 2 (Th2) effector cells that secrete interleukin (IL)-4, IL-5, and IL-13 and induce isotype switching of B lymphocytes (B cells) for IgE synthesis. The sensitization phase is completed by the binding of allergen-specific IgE to the high-affinity FcεRI of mast cells and basophils. When the immune system reencounters the allergen, the effector phase is activated. In this case, the allergen binds to surface-bound IgE and cross-links two FcεRI receptors on mast cells or basophils. This results in the release of chemical mediators, such as histamine and prostaglandin, which induce the typical allergic symptoms (Anvari et al. 2019).

5.2 The Role of the Gut Microbiome in Allergic Diseases

The human gut is inhabited by approximately 40 trillion microorganisms that perform important functions for the host, such as preventing pathogen colonization, detoxifying bile acids, metabolizing nondigestible carbohydrates, and producing key metabolites essential for human health (Mörbe et al. 2021). In addition to these beneficial bacteria, the human gut is also exposed to an array of potential food allergens and harmful microbes. In order to defend itself effectively, it contains a variety of immune compartments and immune cells that work together in a fine-tuned manner (Brown and Esterházy 2021). Gut microbes interact with the immune system through various mechanisms, including receptors for microbial ligands. Binding of such receptors causes the release of cytokines that drive immune responses and lead to either immune tolerance or host defense. Maintaining a balance between immune tolerance and host defense is vital for gut homeostasis. Conversely, disruption of this balance leads to dysregulation of the immune system and consequently to the development of various diseases, including allergies.

Intestinal immune compartments are generally divided into inductive sites, where adaptive immune cells undergo initial activation and differentiation, and effector sites, where primed adaptive immune cells are located and maintained to enhance barrier integrity and immune defense (Brandtzaeg et al. 2008). Gut-associated lymphoid tissue (GALT) structures, such as Peyer's patches (PPs) of the distal ileum, mesenteric lymph nodes (MLNs), and isolated lymphoid follicles (ILFs), belong to inductive compartments and represent the sites where antigens are taken up and presented to T cells and B cells (Brandtzaeg et al. 2008). In the case of food allergens, suppression or tolerance is the predominant response. The process by which the immune system is educated to prevent sensitization to ingested food antigens is poorly understood. However, there is evidence that the gut microbiome plays a key role in this process. Oral tolerance is defined as the induction of mucosal and systemic non-responsiveness to ingested food antigens (Iweala and Nagler 2006). According to the accepted theory, food antigens that cross the epithelial barrier are presented to naïve T cells by specialized populations of DCs in the presence of retinoic acid and transforming growth factor beta (TGF-β). This leads to the robust induction of antigen-specific $FOXP3^+$ regulatory T cells (Tregs), which in turn inhibit naïve T cell activation and differentiation into different Th types (Wesemann and Nagler 2016). Regulatory B cells (Bregs) also contribute to oral tolerance by producing specific IgG4, which further suppresses Th2 immune responses (Noh and Lee 2011).

Experimental research has revealed that commensal bacteria drive the host toward the development of immune tolerance. Indeed, it has been shown that germ-free

(GF) mice, which are typically Th2-skewed, cannot develop oral tolerance to food antigens (Rodriguez et al. 2011). In addition, when GF mice are colonized with gut bacteria from sensitized susceptible mice, food allergy susceptibility is also transmitted (Noval Rivas et al. 2013). Instead, colonization with gut bacteria from healthy infants protects them from food allergies (Feehley et al. 2019). In particular, colonization with certain bacterial strains, such as *Bifidobacterium* and *Clostridium* species, has been shown to promote the induction of mucosal Tregs and thus exert a protective effect against food allergen sensitization (Lyons et al. 2010; Atarashi et al. 2011). Furthermore, there is evidence that commensal bacteria modulate the immune system by secreting metabolites such as short-chain fatty acids (SCFAs). Several studies demonstrate that SCFAs promote the regulatory activity of dendritic cells and lead to the induction of mucosal and peripheral Tregs (Furusawa et al. 2013; Smith et al. 2013; Arpaia et al. 2013). In addition to the induction of antigen-specific Tregs, studies have denoted that the maintenance of oral tolerance requires a protective epithelial barrier response mediated by a class of commensal mucosa–associated bacteria (Stefka et al. 2014; Cao et al. 2014). For example, certain *Clostridium* species have been shown to activate innate lymphoid cells to secrete IL-22, which enhances epithelial barrier integrity and reduces intestinal permeability (Stefka et al. 2014). Another factor that influences the homeostatic regulation of the intestinal mucosal epithelium is the production of secretory IgA (sIgA). The latter is the predominant immunoglobulin on mucosal surfaces, and its primary role, referred to as immune exclusion, is to prevent allergens and pathogens from crossing the mucosal barrier and spreading to the systemic compartment (Corthésy 2013). Indeed, reduced fecal IgA levels have been linked to food allergen sensitization and increased risk of anaphylaxis in mice (Smeekens et al. 2020). However, IgA may also play a role in shaping the composition of the gut microbiome and in regulating the establishment of its reciprocal relationship with its host. In fact, IgA maintains the balance between gut microbes by promoting the retention of beneficial species and the removal of opportunistic pathogens, which in turn limits the absorption of pathogens and toxins. On the other hand, there is evidence that IgA secretion is partly regulated by the gut microbiome and that specific bacterial species can induce its synthesis to varying degrees (León and Francino 2022).

In line with these data, human studies also show that an altered gut microbiome, in terms of both composition and metabolic function, is associated with the development of allergic diseases, particularly food allergy (Goldberg et al. 2020; Fazlollahi et al. 2018; Savage et al. 2018). A recent cohort study showed different taxonomic and metabolomic profiles in healthy and food-allergic twins, highlighting the critical role of the gut microbiome in the development of such disease (Bao et al. 2021). Although the specific gut bacteria involved in the development of food allergy and the ways in which microbes positively or negatively influence the tolerogenic processes that prevent food allergy are still unclear, current research has made significant progress in this area. A recent study conducted by Goldberg et al. showed that the gut microbiome of children with IgE-mediated food allergy is characterized by reduced gut diversity and lower abundance of specific commensal bacteria such as *Bifidobacterium* and *Prevotella copri.* These children also have a different metabolic profile characterized by reduced levels of SCFAs (Goldberg et al. 2020). Children with non-IgE-mediated food allergy also have an altered gut microbiome composition that mainly involves the abundance of *Bacteroides*, *Alistipes*, *Rhodococcus*, *Clostridium* XVIII, Lachnospiraceae, Nocardiaceae, and Erysipelotrichia (Berni Canani et al. 2018; Díaz et al. 2018; Ling et al. 2014). Besides food allergy, gut dysbiosis also occurs in other allergic diseases, such as asthma and atopic dermatitis (Barcik et al. 2020; Fang et al. 2021). Indeed, De Filippis et al. showed that the gut microbiome of children with food or respiratory allergies is characterized by increased abundance of *Ruminococcus gnavus* and *Faecalibacterium prausnitzii*, as well as depletion of *Bifidobacterium longum*, *Bacteroides dorei*, *Bacteroides vulgatus*, and other fiber-degrading taxa. In addition, the metagenomic analysis showed that the allergic children have a pro-inflammatory potential, characterized by enrichment of genes involved in bacterial lipopolysaccharide and urease synthesis (De Filippis et al. 2021) (Fig. 5.1).

5.3 Gut Microbiome Succession Process During Early Life

Interactions between the gut microbiome and the host immune system are established during infancy and early childhood. Under normal conditions, these interactions lead to the induction of immune tolerance, which in turn facilitates the continued proper development of intestinal microbial colonization. In contrast, perturbations in early interactions can have long-term consequences and lead to the development of various diseases, including allergies, later in life (Gensollen et al. 2016; Zheng et al. 2020). The initial colonization process is influenced by several factors, such as genetics, environmental factors, maternal diet, mode of delivery, gestational age, early exposure to antibiotics, and the type of infant feeding (i.e., formula feeding or breastfeeding) (Moriki et al. 2022). Therefore, the period from conception to 2 years of age is critical for the development of both the gut microbiome and the immune system and is referred to as the "window of opportunity" (Robertson et al. 2019).

Despite the widely held theory that the fetal gastrointestinal tract is sterile, recent data indicate that the colonization process may begin in utero during the last trimester of pregnancy. Studies have denoted that maternal gut bacteria can be transmitted to the MLNs and then, through the systemic circulation and the placenta, to the fetus (Jiménez et al. 2008; Moles et al. 2013; Perez et al. 2007). Therefore, maternal nutrition during pregnancy may have long-term effects on the gut colonization process. Indeed, in a recent study, a marked relative depletion of *Bacteroides*, which persisted up to 6 weeks of age, was observed in neonates that were exposed to a high-fat maternal diet during gestation (Chu et al. 2016). However, the gut is mainly exposed to microbes at the time of birth, and thus, the mode of delivery plays an important role in the succession process. Neonates born vaginally are exposed to the vaginal and fecal microbes of their mothers, and their microbiome is dominated by genera such as *Lactobacillus*, *Prevotella*, and *Sneathia*. In contrast, neonates born via cesarean section are predominantly exposed to skin microbes and, thus, harbor

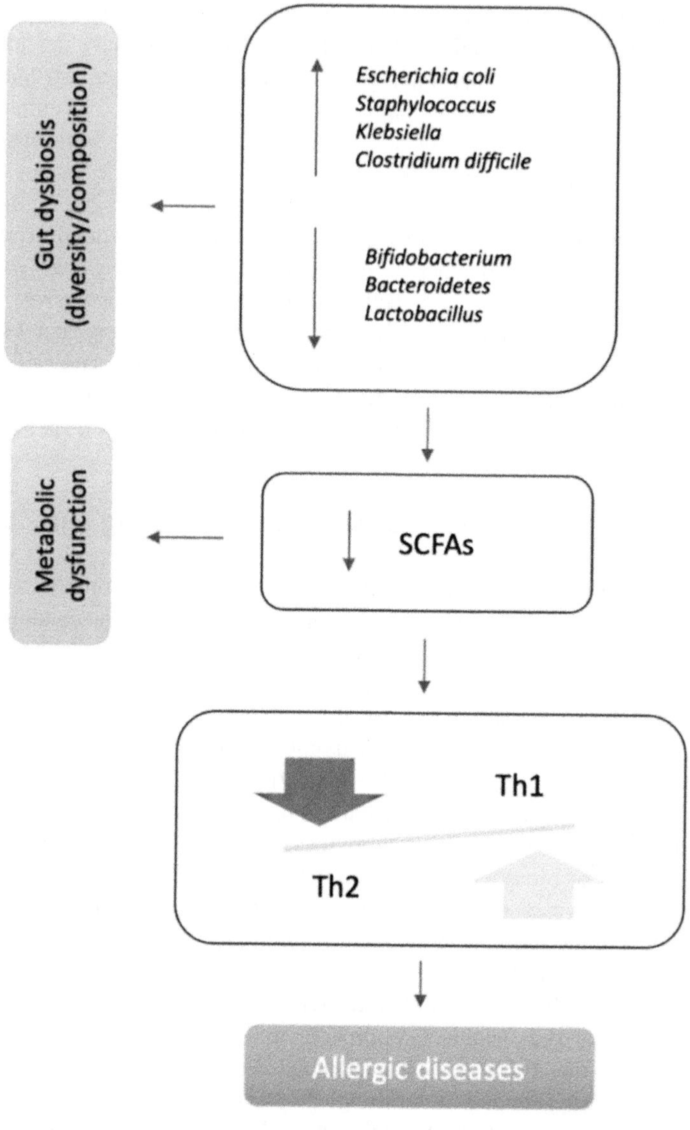

Fig. 5.1 Gut dysbiosis and metabolic dysfunction trigger immune responses that lead to allergic diseases

bacteria such as *Staphylococcus*, *Corynebacterium*, and *Propionibacterium* (Dominguez-Bello et al. 2010). Furthermore, it has been shown that neonates born via cesarean section undergo delayed colonization by *Bifidobacterium* and *Bacteroides* and often have increased levels of *Clostridium difficile* (Azad et al. 2013; Biasucci et al. 2008).

Immediately after birth, newborns are exposed to an avalanche of microbes. As a result, the gut microbiome changes in composition and increases in diversity throughout infancy (Vallès et al. 2012, 2014). The most important events that change the gut microbiome during this period are the introduction of solid foods into the infant's diet and the cessation of breastfeeding (Favier et al. 2002; Palmer et al. 2007). The gut microbiome of breastfed infants is more abundant in commensal microbes, such as *Bifidobacterium* and *Lactobacillus*, compared to that of formula-fed infants, whose microbiome is more diverse and characterized by an abundance of *Escherichia coli*, *Clostridium*, and *Bacteroides* (Levin et al. 2016). These commensals are known to have immunomodulatory properties, and therefore, breastfeeding protects against immune system disorders such as allergies. After the introduction of solid foods into the infant's diet, the gut microbiome converges toward an adult-like microbiome, characterized by greater diversity and richness and dominated by Firmicutes and Bacteroidetes (Bäckhed et al. 2015). In terms of functional repertoire, the gut microbiome of neonates and young infants is enriched in genes that regulate the uptake and degradation of sugars found in breast milk, such as lactose and human milk oligosaccharides (HMOs). Upon weaning, the gut microbiome functionally matures, and there is a decrease in the relative abundance of genes involved in HMOs' breakdown and an enrichment in genes encoding the utilization of a wider range of carbohydrates, including complex sugars and starch (Rey-Mariño and Francino 2022). Nevertheless, it has been shown that sustained breastfeeding delays the development of an adult-like microbiome and has a strong effect on the succession process (Levin et al. 2016; Bäckhed et al. 2015).

In addition to the mode of delivery and type of infant feeding, other factors that influence the succession process during the neonatal period and early infancy include gestational age and exposure to antibiotics. Early antibiotic administration has been associated with reduced gut diversity and specific taxonomic changes, such as a marked increase in the Proteobacteria phylum and decreased levels of *Bifidobacterium* (Tanaka et al. 2009; Fouhy et al. 2012). It is worth noting that neonates whose mothers received antibiotics before birth showed similar but weaker alterations in their gut microbiome (Tanaka et al. 2009; Shao et al. 2019). With regard to gestational age, studies have revealed that the gut microbiome of preterm infants is characterized by reduced gut diversity and lower levels of *Bifidobacterium*, mainly due to the lack or delay of enteral feeding and administration of antibiotics (Magne et al. 2006; Arboleya et al. 2012; Jacquot et al. 2011) (Fig. 5.2).

5.4 Effects of Early Gut Microbiome on Immune System Development

The high incidence of food allergy in infants and young children demonstrates the inadequate capacity to develop oral tolerance. Indeed, infants and young children are vulnerable to infections and immune-mediated disorders, such as allergies, due to poor gut immune function (Battersby and Gibbons 2013). Mucosal immunity will mature postnatally to evolve, ideally, into regulatory immune responses against

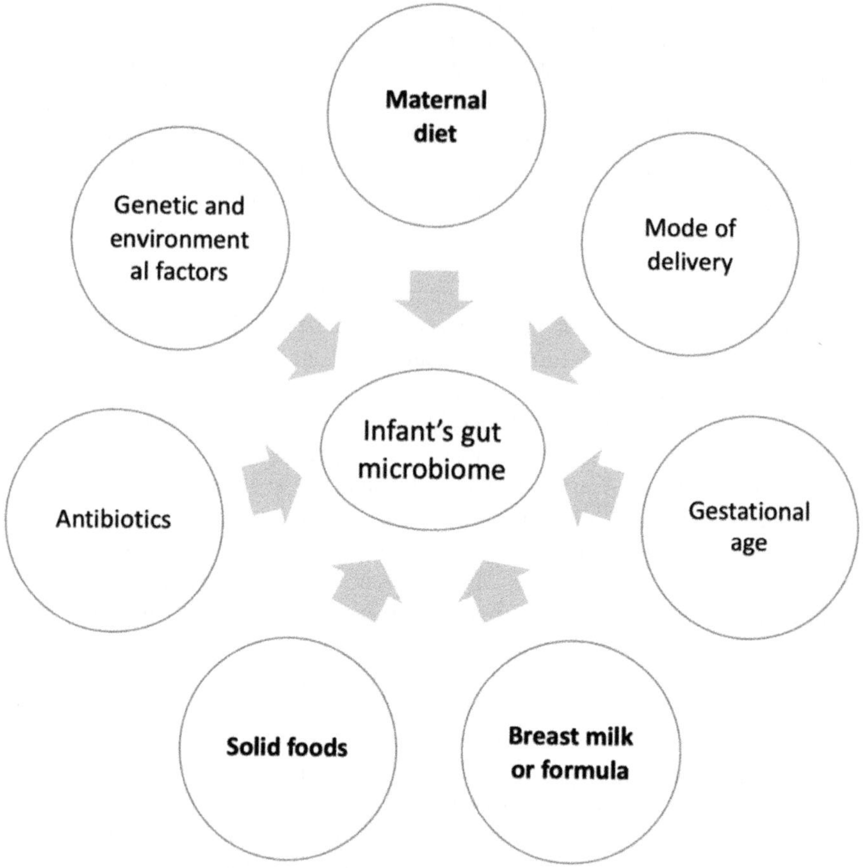

Fig. 5.2 Factors affecting the succession process of the infant's gut microbiome

dietary antigens and nonpathogenic microbes and into inflammatory responses against pathogens. Along with the maturation of the immune system, the gut will evolve from an almost sterile environment at birth to a very simple and highly variable microbiome in the first months of life, which will eventually become dense, complex, and stable at around 2–3 years of age. Therefore, it follows that the maturation of the immune system is likely to be directly and/or indirectly influenced by the gut colonization process. Interactions between gut microbes and the immune system are established during early life and may have long-term consequences. In fact, recent research has begun to delineate a critical period, referred to as the "window of opportunity," during which perturbations can lead to persistent and, in some cases, irreversible abnormalities in the formation and training of certain immune subgroups.

Emerging evidence suggests that commensal microbes have profound effects on the structural and functional development of the immune system. Signals derived from gut microbiome sensing are critical for the development of the intestinal mucosa and GALT structures, such as PPs, MLNs, and ILFs (Maynard et al. 2012). This is supported by studies in GF animals that show significant abnormalities in the development of both primary and secondary lymphoid organs. ILFs are scarcely present in the small intestine where the microbiome is not dense, compared to the large intestine, suggesting that interactions with specific microbial communities directly shape the development of these structures (Macpherson and Harris 2004). In addition, GF animals have hypoplastic PPs and MLNs, which can be reversed within weeks of colonization with gut bacteria (Macpherson and Harris 2004; Baptista et al. 2013). These structural abnormalities are also associated with functional abnormalities such as a reduced frequency of $CD4^+$ and $CD8^+$ intestinal T cell subsets, as well as a reduced number of intraepithelial lymphocytes expressing the T cell receptor (TCR) αβ. Furthermore, it has been shown that GF animals have a general defect in the production of IgA and IgG1 antibodies (Gensollen et al. 2016), as well as elevated total serum IgE levels after weaning, due to increased B cell isotype class switching to IgE in PPs and MLNs (Cahenzli et al. 2013). These abnormalities can be attributed to the lack of specific bacterial populations that are able to modulate various immune responses, including those leading to IgA secretion; Th1, Th2, and Th17 cell differentiation; and expansion of $FOXP3^+$ Treg in the colon (Rooks and Garrett 2016). For example, it has been shown that monocolonization of GF mice with *Bacteroides fragilis* corrects the imbalance between Th1 and Th2 cells and promotes Treg differentiation (Mazmanian et al. 2005). Furthermore, there is evidence that segmented filamentous bacteria (SFB) play an important role in intestinal Th cell responses in mice, particularly in the development of Th17 cells, which are critical for mucosal barrier integrity (Gaboriau-Routhiau et al. 2009; Ivanov et al. 2009). SFBs are also potent inducers of sIgA and probably have a unique role in the postnatal maturation of gut immune function (Gaboriau-Routhiau et al. 2009). Finally, colonization with certain *Clostridium* species has been shown to promote the induction of Tregs in the colon of GF mice (Feehley et al. 2019; Atarashi et al. 2011). Nevertheless, it seems that the interplay between gut microbes and the immune system is more complex and that a diverse gut microbiome is required to develop a fully mature immune system.

Gut microbes exert their effects on the immune system both directly through signaling via microorganism-associated molecular patterns (MAMPs) and indirectly through the production of several metabolites. MAMPs are recognized by a variety of germline-encoded PRRs, such as Toll-like receptors (TLRs), C-type lectin receptors (CLRs), and nucleotide-binding oligomerization domain (NOD)-like receptors (NLRs) that are produced by epithelial and immune cells in the gut (Maynard et al. 2012). One of the most well-studied MAMPs is polysaccharide A (PSA). PSA is produced by *Bacteroides fragilis* and has been shown to correct the imbalance between Th1 and Th2 cells. In addition, it promotes Treg differentiation and MyD88 signaling, which is involved in epithelial repair and antimicrobial peptide secretion (Rooks and Garrett 2016; Wang et al. 2015). SCFAs are among the most important metabolites with immunomodulatory properties. They include acetate, butyrate, and propionate and are produced by anaerobic bacteria, such as clostridia and bifidobacteria, during the fermentation of non-digested carbohydrates that reach the colon. SCFAs act both epigenetically by directly inhibiting histone deacetylases (HDACs) that alter gene expression and through G protein-coupled receptors (GPCRs), such as GPR43, GPR41, and GPR109A, which are present in epithelial and immune cells (Rooks and Garrett 2016). SCFAs enhance immune tolerance by promoting the induction of Tregs in the intestinal environment (Smith et al. 2013; Arpaia et al. 2013), reduce NF-kB activity, and inhibit the release of pro-inflammatory cytokines such as IL-12, tumor necrosis factor alpha (TNF-α), IL-1b, and nitric oxide from macrophages and neutrophils (Ni et al. 2010; Säemann et al. 2000; Segain et al. 2000). SCFAs also promote mucus secretion and IgA production and maintain gut integrity by accelerating the assembly of tight junction (TJ) proteins, such as ZO-1 and occludin (Balda and Matter 2008). Other important metabolites include kynurenine, 3-hydroxykynurenine, and xanthurenic acid, which are produced by commensal bacteria, such as *Lactobacillus*, during the metabolism of tryptophan. Such metabolites bind the aryl hydrocarbon receptor (AhR), a nuclear receptor that metabolizes toxic compounds, and are considered key regulators of Tregs and Th17 cell function (Quintana et al. 2008). Moreover, there is evidence that AhR plays a key role in mast cell and dendritic cell functions (Sibilano et al. 2012; Schulz et al. 2013).

In addition to which microbes are necessary to restore a fully mature immune system, another question that arises is the time frame for the restoration of immune deficiencies. Although the majority of abnormalities in GF animals are not age-dependent and can be corrected by introducing commensal microbes at any age, the ability to repair some cellular defects that develop in the absence of gut microbes is limited to a critical period in early life (Gensollen et al. 2016). For example, elevated total serum IgE levels in GF mice can only be normalized if colonization occurs before 4 weeks (Cahenzli et al. 2013). Furthermore, the accumulation of invariant natural killer T (iNKT) cells in the lung of GF mice, which leads to airway hyperresponsiveness and asthma susceptibility, can only be restored by colonization with standard gut bacteria during the first 2 weeks of life (Gensollen et al. 2016). It therefore follows that there is a critical period for the development of the immune system that determines lifelong health.

Consistent with these data, perturbations in gut microbiome composition and function during the critical period of immune system regulation have been associated with the development of allergic diseases later in life in several human studies. A large birth cohort study of 411 high-risk infants showed that gut bacterial diversity at 1 and 12 months of age is inversely associated with the risk of allergic sensitization and the development of allergic rhinitis during the first 6 years of life (Bisgaard et al. 2011). These results are also confirmed by other studies, in which reduced gut diversity in early infancy has consistently been associated with food allergen sensitization and an increased risk of allergic diseases later in childhood (Azad et al. 2015; Abrahamsson et al. 2012; Sjögren et al. 2009). However, when trying to explain the role of the gut microbiome in the development of allergic diseases, specific taxonomic and metabolic changes are considered more important than microbial diversity. For instance, a higher Enterobacteriaceae/Bacteroidaceae

(E/B) ratio, which is indicative of an immature gut microbiome, early in life has been associated with an increased risk of later sensitization to food allergens (Azad et al. 2015). In addition, a large longitudinal study of 319 infants showed that transient gut dysbiosis and metabolic dysfunction during early life are associated with an increased risk of atopy and wheezing. In particular, a selective reduction in the relative abundance of specific genera such as *Lachnospira, Veillonella, Faecalibacterium,* and *Rothia*, accompanied by reduced fecal acetate levels, was observed at 3 months of age in infants who developed wheezing and positive skin prick tests (SPTs) at 12 months. However, these alterations were not evident in fecal samples obtained at 12 months of age, suggesting that the first 100 days of life offer a "window of opportunity" for immune regulation (Arrieta et al. 2015). Another randomized controlled trial also showed that infants at risk for asthma have a distinct, but plastic early-life gut microbial development that can be partially restored after oral supplementation with probiotics such as *Lactobacillus rhamnosus* GG (LGG). In particular, the meconium of high-risk infants was enriched in Enterobacteriaceae and Bacteroidaceae and depleted in multiple genera including *Akkermansia, Faecalibacterium,* and *Rothia.* In addition, these infants showed delayed diversification of the gut microbiome, which was predominantly glycolytic and depleted of a range of anti-inflammatory lipids at 6 months of age. These deficits were partially restored in infants supplemented with LGG, but this effect was lost 6 months after the cessation of supplementation (Durack et al. 2018).

These data suggest that the succession process of the gut microbiome during early life plays an important role in the development of allergic diseases, and therefore, its modification during this critical period may be the key to the prevention and treatment of such diseases. However, more research is needed to better define the factors that influence the succession process and how gut microbes interact with each other and with the immune system, leading to the development of immune tolerance.

5.5 Nutrition and Allergic Diseases

5.5.1 Maternal Nutrition and Allergic Diseases in Offspring

Most allergic diseases begin in infancy and early childhood. Maternal diet during pregnancy, and eventually during lactation, is the earliest nutritional factor that may influence allergic outcomes in offspring. However, most epidemiological studies have failed to report consistent associations (Netting et al. 2014). This is not only because dietary patterns are often loosely defined in terms of their content of nutrients, vitamins, and antioxidants but also because there are other confounding factors such as parental atopic status, mode of delivery, number of siblings, exposure to common allergens, maternal and early-life viral infections, and exposure to tobacco smoke (Julia et al. 2015). Despite the discrepancies, maternal diets rich in fruits, vegetables, fish, and vitamin D–containing foods, such as the Mediterranean diet, have been shown to provide consistent benefits on allergic sensitization and the development of allergic diseases in offspring (Netting et al. 2014). On the contrary, there is some evidence that maternal diets high in fat and sugar increase the risk of allergic diseases in offspring (Netting et al. 2014; Gupta et al. 2022).

Studies focusing on specific foods or nutrients have led to more consistent conclusions. For example, maternal supplementation during pregnancy with fish oil, which is rich in omega-3 polyunsaturated fatty acids (PUFAs), has been correlated with a lower risk of allergen sensitization and atopic dermatitis in children (Dunstan et al. 2003). Moreover, according to a large Danish cohort, maternal intake of vitamin A and vitamin E during pregnancy may have a protective effect against allergic rhinitis in children (Maslova et al. 2014). Finally, a recent post hoc analysis showed that vitamin D supplementation during pregnancy is associated with a significant reduction in asthma between 3 and 6 years of age, as well as reduced total serum IgE levels and improved lung function in offspring (Weiss et al. 2024). Analyses of other antioxidant molecules and nutrients do not indicate further associations between childhood allergies and their status during pregnancy. Besides maternal dietary intake, a recent prospective cohort study concluded that higher consumption of food allergens, such as milk, peanuts, and wheat, during pregnancy is associated with a reduced risk of allergy and asthma in offspring at the age of 8 years (Bunyavanich et al. 2014).

Nutrition has a significant impact on both the composition and metabolic function of the gut microbiome. A high-fiber maternal diet has been linked to increased diversity and richness of the gut microbiome during pregnancy, as well as increased abundance in certain bacterial taxa such as *Holdemania, Roseburia, Lachnospira,* and *Coprococcus* (Barrett et al. 2018; Gomez-Arango et al. 2018; Röytiö et al. 2017). These bacteria ferment fiber into SCFAs in the mother's large intestine, which in turn cross the placenta and enter the fetal circulation. SCFAs are known to have potent local and systemic immune effects and, thus, modulate allergic outcomes in offspring. Indeed, a high-fiber maternal diet has been associated with a reduced airway inflammatory response to ovalbumin challenge in offspring in an experimental model (Thorburn et al. 2015). Therefore, there is some evidence that a high-fiber maternal diet protects the offspring against asthma. The composition of the gut microbiome during pregnancy has also been associated with the development of food allergy in offspring. In particular, it has been demonstrated that the maternal carriage of *Prevotella copri* may protect against the development of IgE-mediated food allergy at 1 year of age (Vuillermin et al. 2020). It has also been suggested that this is achieved through the production of succinate, which promotes bone marrow myelopoiesis of dendritic cell progenitors in the embryo (Gao et al. 2021).

5.5.2 Early-Life Nutrition and Allergic Diseases

5.5.2.1 Breastfeeding

Milk is the primary source of energy and nutrition during infancy. The infant's feeding pattern (i.e., breastfeeding or formula feeding) has been associated with the risk of infections, inflammatory diseases, and autoimmune disorders both in infancy and later in life (Hosea Blewett et al. 2008). In particular, breastfeeding has been shown to have a protective effect against allergies, especially in infants with a family history of atopic diseases or other risk factors (van Odijk et al. 2003). Breast milk contains hundreds of nonpathogenic bacteria, essential micro- and macronutrients, and various bioactive factors that influence important biological processes, including intestinal colonization and maturation of the immune system. Such factors include immunoglobulins, antimicrobial peptides, anti-inflammatory cytokines, growth factors, and hormones (Ballard and Morrow 2013). Therefore, breast milk promotes the colonization of the gut by commensal microbes, such as *Bifidobacterium* and *Lactobacillus*, which in turn educate the newborn's immature immune system to distinguish between these commensal bacteria and those that may be harmful (Vidal and Donnet-Hughes 2008).

HMOs are among the most important bioactive factors in breast milk. They act as prebiotics and promote the development of a healthy gut microbiome composition dominated by commensal bacteria, such as bifidobacteria, which metabolize HMOs into SCFAs and favor the regulation of the immune system. They also promote the maturation of the intestinal mucosa and epithelial barrier and prevent pathogens from adhering to the epithelial cells. Furthermore, HMOs modulate various immune cells and PRRs, as well as signaling pathways associated with GALT maturation and cytokine networks that regulate the balance between Th1 and Th2 immune responses (Sekerel et al. 2021). Indeed, experimental studies in mice have revealed that HMOs, in particular 2′-fucosyllactose (2′-FL) and 6′-sialyllactose (6′-SL), induce IL-10^+ Tregs and promote the development of oral tolerance (Castillo-Courtade et al. 2015). The beneficial effects of HMOs on allergies are also supported by studies in infants. Specifically, HMOs have been shown to protect infants from food allergen sensitization (Miliku et al. 2018), atopic dermatitis (Sprenger et al. 2017), and cow's milk allergy (Seppo et al. 2017).

Breast milk also provides infants with passive mucosal immunity through sIgA, and other non-antibody and antimicrobial factors such as lactoferrin and lysozyme. In addition to passive mucosal immunity, these factors promote the establishment of a healthy gut microbiome and, consequently, the proper development of the immune system (Turfkruyer and Verhasselt 2015). Furthermore, breast milk contains several immunomodulatory factors such as vitamin A and omega-3 PUFAs. The latter are known to have anti-inflammatory properties and enhance the integrity of the intestinal barrier. Indeed, in a mouse model, an increase in the amount of omega-3 PUFAs in breast milk was associated with a greater capacity to induce oral tolerance to ovalbumin (Korotkova et al. 2004). These results are further supported by studies in infants, which show that increased levels of omega-3 PUFAs are associated with reduced allergen-specific Th2 responses and protect against allergic sensitization during early life (Duchén et al. 1998; D'Vaz et al. 2012).

5.5.2.2 Introduction of Solid Foods

Solid foods are usually introduced into the infant's diet between 4 and 6 months of age. Although it has previously been suggested that the delayed introduction of solid foods could prevent the development of allergic diseases, recent data do not support this hypothesis (Greer et al. 2008). In contrast, the PASTURE/EFRAIM study, which is a cohort including children from rural areas in 5 European countries, showed that the introduction of an increased variety of complementary foods in the first year of life is inversely associated with the development of asthma, allergic rhinitis, food allergy and atopic sensitisation by the age of 6 years (Roduit et al. 2014). In addition, although not recommended due to possible contamination with pathogens, consumption of unprocessed milk during the first year of life has been negatively associated with asthma, allergic rhinitis, and atopy (Riedler et al. 2001). Interestingly, further research identified whey proteins, such as α-lactalbumin, β-lactoglobulin, and bovine serum albumin, as the components of unprocessed milk that are associated with a reduced risk of asthma, but not of atopy (Loss et al. 2011).

Increasing evidence suggests that exposure to diet-derived antigens is critical for the development of immune tolerance. After weaning, food antigens are necessary and sufficient to activate memory $CD4^+$ T cells and peripherally induced Tregs to colonize the lamina propria in the gut and control the Th2 immune response. In addition, exposure to food antigens is essential for the production of IgA and IgG, which also contribute to the induction of oral tolerance (Menezes et al. 2003). Indeed, studies revealed that early egg introduction, between 4 and 6 months of age, can prevent egg allergy. Similarly, early introduction of peanuts, between 4 and 11 months of age, can prevent peanut allergy (Palmer et al. 2017). However, this is not the case with the early introduction of other food allergens, such as fish and cow's milk. Therefore, these data suggest that the promotion of oral tolerance through antigen exposure during the critical period of immune regulation is possible, but other additional factors may also play a critical role.

5.5.3 Nutrition in Childhood and Allergic Diseases

After the introduction of complementary foods, the pattern of dietary diversification will undoubtedly affect the establishment of the child's gut microbiome and, consequently, the development of allergic diseases. Several studies have denoted that adherence to healthy dietary patterns, such as the Mediterranean diet, during childhood protects children from allergic diseases, especially asthma (Chatzi et al. 2007; Garcia-Marcos et al. 2013; Koumpagioti et al. 2022). This is probably due to the high content of certain nutrients and food components, such as fiber and unsaturated fatty acids, which strongly influence the composition and metabolic function of the gut microbiome.

Dietary fibers are nondigestible carbohydrates that are fermented by anaerobic gut bacteria into SCFAs. Adherence to the Mediterranean diet, which is rich in fiber, has been shown to increase the abundance of *Prevotella* and other fiber-degrading bacteria, such as *Lachnospira* (De Filippis et al. 2016). These taxonomic changes are also accompanied by increased levels of SCFAs, which are known to play a key role in preventing the development of allergic diseases. Indeed, in a cohort study, high fecal butyrate levels at 1 year of age were significantly associated with less atopic sensitization to food and inhalant allergens, as well as a lower risk of asthma, food allergy, and allergic rhinitis between the ages of 3 and 6 years (Roduit et al. 2019). Several experimental studies have shown that SCFAs protect against allergic airway inflammation and food allergy. In particular, propionate activates GPR41 receptors, which in turn promote the development of inflammatory monocytes. Upon migration to the lungs, monocytes are converted into DCs with a reduced ability to promote the differentiation of naïve T cells into Th2 cells (Trompette et al. 2014). Nevertheless, there is evidence that SCFAs also exert their beneficial effects through other GPCRs, such as GRP43 (Maslowski et al. 2009), as well as through the inhibition of HDAC activity. Indeed, trichostatin A, an HDAC inhibitor, reduced allergic airway inflammation in a mouse model, suggesting that the protective effects of SCFAs against asthma may also be exerted through the regulation of HDAC activity (Choi et al. 2005). Finally, SCFAs induce colonic Tregs, both directly through the inhibition of HDAC activity and indirectly through the activation of GPR109A, resulting in oral tolerance (Arpaia et al. 2013; Singh et al. 2014). In contrast, adherence to Western diets, characterized by high consumption of fats and sugars and low-fiber intake, has been associated with lower levels of SCFAs and an increased risk of allergic diseases (De Filippo et al. 2010).

The Mediterranean diet is also rich in omega-3 PUFAs. The latter are key fatty acids with anti-inflammatory properties. They include a-linolenic acid (ALA), eicosapentaenoic acid (EPA), and docosahexaenoic acid (DHA) and have been shown to have beneficial effects on allergies. Experimental studies show that omega-3 PUFA supplementation protects mice from food allergy and atopic dermatitis (de Matos et al. 2012; Weise et al. 2013). In addition, there is evidence that Fat-1 transgenic mice, which have increased levels of omega-3 PUFAs, have lower allergic inflammation (Bilal et al. 2011). Omega-3 PUFAs act on several immune cells that are involved in the allergic response. Specifically, DHA suppresses Th2 immune responses both directly through targeting IL-13 (MacLean et al. 2013) and indirectly through the induction of Tregs (Han et al. 2012). DHA also inhibits IgE isotype switching in B cells by interfering with CD40 and IL-4 receptor signaling (Weise et al. 2011). Finally, EPA and DHA suppress the production of reactive oxygen species (ROS) and the release of IL-4 and IL-13 by human mast cells in vitro (van den Elsen et al. 2013). A recent cross-sectional study conducted in adults concluded that a higher omega-3 ratio is associated with improved asthma control and a lower need for inhaled corticosteroids (Stoodley et al. 2019). Similarly, omega-3 PUFAs have been shown to have beneficial effects on asthma in children and adolescents (Zhang et al. 2023). Conversely, the typical Western diet is rich in omega-6 PUFAs, such as linoleic acid, which are considered pro-inflammatory. Therefore, maintaining a low omega-6 to omega-3 PUFA ratio is important for reducing the incidence of asthma and other allergic diseases (DiNicolantonio and O'Keefe 2021).

5.6 Conclusions

The succession process of the gut microbiome during early life has a strong impact on the development of the immune system. In particular, mucosal immunity is strongly influenced by antigens derived from the gut microbiome and by food metabolites produced by gut bacteria, such as SCFAs. Disruption of this process can have long-term consequences and can contribute to the development of allergic diseases. Nutrition is one of the most important factors affecting the gut microbiome and is therefore vital for the ontogeny and function of the immune system. Before weaning, breast milk accumulates all the exogenous factors necessary for its development. Antigens derived from solid foods, vitamins, lipids, and food metabolites will then continue to modulate immune function and determine long-term susceptibility to local and systemic immune-mediated diseases. As a consequence, dietary interventions are a potential approach that could safely contribute to the prevention or treatment of allergic diseases. However, more research is needed to better understand the underlying processes by which different dietary components promote or inhibit the development of allergic diseases in order to design personalized diets for allergic patients.

References

Abrahamsson TR, Jakobsson HE, Andersson AF, Björkstén B, Engstrand L, Jenmalm MC (2012) Low diversity of the gut microbiota in infants with atopic eczema. J Allergy Clin Immunol 129(2):434–40, 40.e1–2

Akdis CA (2021) Does the epithelial barrier hypothesis explain the increase in allergy, autoimmunity and other chronic conditions? Nat Rev Immunol 21(11):739–751

Anvari S, Miller J, Yeh CY, Davis CM (2019) IgE-mediated food allergy. Clin Rev Allergy Immunol 57(2):244–260

Arboleya S, Solís G, Fernández N, de los Reyes-Gavilán CG, Gueimonde M (2012) Facultative to strict anaerobes ratio in the preterm infant microbiota: a target for intervention? Gut Microbes 3(6):583–588

Arpaia N, Campbell C, Fan X, Dikiy S, van der Veeken J, deRoos P et al (2013) Metabolites produced by commensal bacteria promote peripheral regulatory T-cell generation. Nature 504(7480):451–455

Arrieta MC, Stiemsma LT, Dimitriu PA, Thorson L, Russell S, Yurist-Doutsch S et al (2015) Early infancy microbial and metabolic alterations affect risk of childhood asthma. Sci Transl Med 7(307):307ra152

Atarashi K, Tanoue T, Shima T, Imaoka A, Kuwahara T, Momose Y et al (2011) Induction of colonic regulatory T cells by indigenous clostridium species. Science 331(6015):337–341

Azad MB, Konya T, Maughan H, Guttman DS, Field CJ, Chari RS et al (2013) Gut microbiota of healthy Canadian infants: profiles by mode of delivery and infant diet at 4 months. CMAJ 185(5):385–394

Azad MB, Konya T, Guttman DS, Field CJ, Sears MR, HayGlass KT et al (2015) Infant gut microbiota and food sensitization: associations in the first year of life. Clin Exp Allergy 45(3): 632–643

Bäckhed F, Roswall J, Peng Y, Feng Q, Jia H, Kovatcheva-Datchary P et al (2015) Dynamics and stabilization of the human gut microbiome during the first year of life. Cell Host Microbe 17(5): 690–703

Balda MS, Matter K (2008) Tight junctions at a glance. J Cell Sci 121(Pt 22):3677–3682

Ballard O, Morrow AL (2013) Human milk composition: nutrients and bioactive factors. Pediatr Clin N Am 60(1):49–74

Bao R, Hesser LA, He Z, Zhou X, Nadeau KC, Nagler CR (2021) Fecal microbiome and metabolome differ in healthy and food-allergic twins. J Clin Invest 131(2):e141935

Baptista AP, Olivier BJ, Goverse G, Greuter M, Knippenberg M, Kusser K et al (2013) Colonic patch and colonic SILT development are independent and differentially regulated events. Mucosal Immunol 6(3):511–521

Barcik W, Boutin RCT, Sokolowska M, Finlay BB (2020) The role of lung and gut microbiota in the pathology of asthma. Immunity 52(2):241–255

Barrett HL, Gomez-Arango LF, Wilkinson SA, McIntyre HD, Callaway LK, Morrison M et al (2018) A vegetarian diet is a major determinant of gut microbiota composition in early pregnancy. Nutrients 10(7):890

Battersby AJ, Gibbons DL (2013) The gut mucosal immune system in the neonatal period. Pediatr Allergy Immunol 24(5):414–421

Berni Canani R, De Filippis F, Nocerino R, Paparo L, Di Scala C, Cosenza L et al (2018) Gut microbiota composition and butyrate production in children affected by non-IgE-mediated cow's milk allergy. Sci Rep 8(1):12500

Biasucci G, Benenati B, Morelli L, Bessi E, Boehm G (2008) Cesarean delivery may affect the early biodiversity of intestinal bacteria. J Nutr 138(9):1796s–1800s

Bilal S, Haworth O, Wu L, Weylandt KH, Levy BD, Kang JX (2011) Fat-1 transgenic mice with elevated omega-3 fatty acids are protected from allergic airway responses. Biochim Biophys Acta 1812(9):1164–1169

Bisgaard H, Li N, Bonnelykke K, Chawes BL, Skov T, Paludan-Müller G et al (2011) Reduced diversity of the intestinal microbiota during infancy is associated with increased risk of allergic disease at school age. J Allergy Clin Immunol 128(3):646–52.e1–5

Brandtzaeg P, Kiyono H, Pabst R, Russell MW (2008) Terminology: nomenclature of mucosa-associated lymphoid tissue. Mucosal Immunol 1(1):31–37

Brown H, Esterházy D (2021) Intestinal immune compartmentalization: implications of tissue specific determinants in health and disease. Mucosal Immunol 14(6):1259–1270

Bunyavanich S, Rifas-Shiman SL, Platts-Mills TA, Workman L, Sordillo JE, Camargo CA Jr et al (2014) Peanut, milk, and wheat intake during pregnancy is associated with reduced allergy and asthma in children. J Allergy Clin Immunol 133(5):1373–1382

Cahenzli J, Köller Y, Wyss M, Geuking MB, McCoy KD (2013) Intestinal microbial diversity during early-life colonization shapes long-term IgE levels. Cell Host Microbe 14(5):559–570

Cao S, Feehley TJ, Nagler CR (2014) The role of commensal bacteria in the regulation of sensitization to food allergens. FEBS Lett 588(22):4258–4266

Castillo-Courtade L, Han S, Lee S, Mian FM, Buck R, Forsythe P (2015) Attenuation of food allergy symptoms following treatment with human milk oligosaccharides in a mouse model. Allergy 70(9):1091–1102

Chatzi L, Apostolaki G, Bibakis I, Skypala I, Bibaki-Liakou V, Tzanakis N et al (2007) Protective effect of fruits, vegetables and the Mediterranean diet on asthma and allergies among children in Crete. Thorax 62(8):677–683

Choi JH, Oh SW, Kang MS, Kwon HJ, Oh GT, Kim DY (2005) Trichostatin a attenuates airway inflammation in mouse asthma model. Clin Exp Allergy 35(1):89–96

Chu DM, Antony KM, Ma J, Prince AL, Showalter L, Moller M et al (2016) The early infant gut microbiome varies in association with a maternal high-fat diet. Genome Med 8(1):77

Corthésy B (2013) Multi-faceted functions of secretory IgA at mucosal surfaces. Front Immunol 4: 185

D'Vaz N, Meldrum SJ, Dunstan JA, Lee-Pullen TF, Metcalfe J, Holt BJ et al (2012) Fish oil supplementation in early infancy modulates developing infant immune responses. Clin Exp Allergy 42(8):1206–1216

De Filippis F, Pellegrini N, Vannini L, Jeffery IB, La Storia A, Laghi L et al (2016) High-level adherence to a Mediterranean diet beneficially impacts the gut microbiota and associated metabolome. Gut 65(11):1812–1821

De Filippis F, Paparo L, Nocerino R, Della Gatta G, Carucci L, Russo R et al (2021) Specific gut microbiome signatures and the associated pro-inflammatory functions are linked to pediatric allergy and acquisition of immune tolerance. Nat Commun 12(1):5958

De Filippo C, Cavalieri D, Di Paola M, Ramazzotti M, Poullet JB, Massart S et al (2010) Impact of diet in shaping gut microbiota revealed by a comparative study in children from Europe and rural Africa. Proc Natl Acad Sci USA 107(33):14691–14696

de Matos OG, Amaral SS, Pereira da Silva PE, Perez DA, Alvarenga DM, Ferreira AV et al (2012) Dietary supplementation with omega-3-PUFA-rich fish oil reduces signs of food allergy in ovalbumin-sensitized mice. Clin Dev Immunol 2012:236564

Díaz M, Guadamuro L, Espinosa-Martos I, Mancabelli L, Jiménez S, Molinos-Norniella C et al (2018) Microbiota and derived parameters in fecal samples of infants with non-IgE cow's milk protein allergy under a restricted diet. Nutrients 10(10):1481

DiNicolantonio JJ, O'Keefe J (2021) The importance of maintaining a low Omega-6/Omega-3 ratio for reducing the risk of autoimmune diseases, asthma, and allergies. Mo Med 118(5):453–459

Dominguez-Bello MG, Costello EK, Contreras M, Magris M, Hidalgo G, Fierer N et al (2010) Delivery mode shapes the acquisition and structure of the initial microbiota across multiple body habitats in newborns. Proc Natl Acad Sci 107(26):11971–11975

Duchén K, Yu G, Björkstén B (1998) Atopic sensitization during the first year of life in relation to long chain polyunsaturated fatty acid levels in human milk. Pediatr Res 44(4):478–484

Dunstan JA, Mori TA, Barden A, Beilin LJ, Taylor AL, Holt PG et al (2003) Fish oil supplementation in pregnancy modifies neonatal allergen-specific immune responses and clinical outcomes in infants at high risk of atopy: a randomized, controlled trial. J Allergy Clin Immunol 112(6): 1178–1184

Durack J, Kimes NE, Lin DL, Rauch M, McKean M, McCauley K et al (2018) Delayed gut microbiota development in high-risk for asthma infants is temporarily modifiable by lactobacillus supplementation. Nat Commun 9(1):707

Fang Z, Li L, Zhang H, Zhao J, Lu W, Chen W (2021) Gut microbiota, probiotics, and their interactions in prevention and treatment of atopic dermatitis: a review. Front Immunol 12: 720393

Favier CF, Vaughan EE, De Vos WM, Akkermans ADL (2002) Molecular monitoring of succession of bacterial communities in human neonates. Appl Environ Microbiol 68(1):219–226

Fazlollahi M, Chun Y, Grishin A, Wood RA, Burks AW, Dawson P et al (2018) Early-life gut microbiome and egg allergy. Allergy 73(7):1515–1524

Feehley T, Plunkett CH, Bao R, Choi Hong SM, Culleen E, Belda-Ferre P et al (2019) Healthy infants harbor intestinal bacteria that protect against food allergy. Nat Med 25(3):448–453

Fouhy F, Guinane CM, Hussey S, Wall R, Ryan CA, Dempsey EM et al (2012) High-throughput sequencing reveals the incomplete, short-term recovery of infant gut microbiota following parenteral antibiotic treatment with ampicillin and gentamicin. Antimicrob Agents Chemother 56(11):5811–5820

Furusawa Y, Obata Y, Fukuda S, Endo TA, Nakato G, Takahashi D et al (2013) Commensal microbe-derived butyrate induces the differentiation of colonic regulatory T cells. Nature 504(7480):446–450

Gaboriau-Routhiau V, Rakotobe S, Lécuyer E, Mulder I, Lan A, Bridonneau C et al (2009) The key role of segmented filamentous bacteria in the coordinated maturation of gut helper T cell responses. Immunity 31(4):677–689

Gao Y, Nanan R, Macia L, Tan J, Sominsky L, Quinn TP et al (2021) The maternal gut microbiome during pregnancy and offspring allergy and asthma. J Allergy Clin Immunol 148(3):669–678

Garcia-Marcos L, Castro-Rodriguez JA, Weinmayr G, Panagiotakos DB, Priftis KN, Nagel G (2013) Influence of Mediterranean diet on asthma in children: a systematic review and meta-analysis. Pediatr Allergy Immunol 24(4):330–338

Gensollen T, Iyer SS, Kasper DL, Blumberg RS (2016) How colonization by microbiota in early life shapes the immune system. Science 352(6285):539–544

Goldberg MR, Mor H, Magid Neriya D, Magzal F, Muller E, Appel MY et al (2020) Microbial signature in IgE-mediated food allergies. Genome Med 12(1):92

Gomez-Arango LF, Barrett HL, Wilkinson SA, Callaway LK, McIntyre HD, Morrison M et al (2018) Low dietary fiber intake increases Collinsella abundance in the gut microbiota of overweight and obese pregnant women. Gut Microbes 9(3):189–201

Greer FR, Sicherer SH, Burks AW (2008) Effects of early nutritional interventions on the development of atopic disease in infants and children: the role of maternal dietary restriction, breastfeeding, timing of introduction of complementary foods, and hydrolyzed formulas. Pediatrics 121(1):183–191

Gupta A, Singh A, Fernando RL, Dharmage SC, Lodge CJ, Waidyatillake NT (2022) The association between sugar intake during pregnancy and allergies in offspring: a systematic review and a meta-analysis of cohort studies. Nutr Rev 80(4):904–918

Han SC, Kang GJ, Ko YJ, Kang HK, Moon SW, Ann YS et al (2012) Fermented fish oil suppresses T helper 1/2 cell response in a mouse model of atopic dermatitis via generation of CD4+-CD25+Foxp3+ T cells. BMC Immunol 13:44

Hosea Blewett HJ, Cicalo MC, Holland CD, Field CJ (2008) The immunological components of human milk. Adv Food Nutr Res 54:45–80

Ivanov II, Atarashi K, Manel N, Brodie EL, Shima T, Karaoz U et al (2009) Induction of intestinal Th17 cells by segmented filamentous bacteria. Cell 139(3):485–498

Iweala OI, Nagler CR (2006) Immune privilege in the gut: the establishment and maintenance of non-responsiveness to dietary antigens and commensal flora. Immunol Rev 213:82–100

Jacquot A, Neveu D, Aujoulat F, Mercier G, Marchandin H, Jumas-Bilak E et al (2011) Dynamics and clinical evolution of bacterial gut microflora in extremely premature patients. J Pediatr 158(3):390–396

Jiménez E, Marín ML, Martín R, Odriozola JM, Olivares M, Xaus J et al (2008) Is meconium from healthy newborns actually sterile? Res Microbiol 159(3):187–193

Julia V, Macia L, Dombrowicz D (2015) The impact of diet on asthma and allergic diseases. Nat Rev Immunol 15(5):308–322

Korotkova M, Telemo E, Yamashiro Y, Hanson LA, Strandvik B (2004) The ratio of n-6 to n-3 fatty acids in maternal diet influences the induction of neonatal immunological tolerance to ovalbumin. Clin Exp Immunol 137(2):237–244

Koumpagioti D, Boutopoulou B, Moriki D, Priftis KN, Douros K (2022) Does adherence to the Mediterranean diet have a protective effect against asthma and allergies in children? A systematic review. Nutrients 14(8):100948

Larsen SB, Cowley CJ, Fuchs E (2020) Epithelial cells: liaisons of immunity. Curr Opin Immunol 62:45–53

León ED, Francino MP (2022) Roles of secretory immunoglobulin a in host-microbiota interactions in the gut ecosystem. Front Microbiol 13:880484

Levin AM, Sitarik AR, Havstad SL, Fujimura KE, Wegienka G, Cassidy-Bushrow AE et al (2016) Joint effects of pregnancy, sociocultural, and environmental factors on early life gut microbiome structure and diversity. Sci Rep 6(1):31775

Li D, Wu M (2021) Pattern recognition receptors in health and diseases. Signal transduction and targeted. Therapy 6(1):291

Ling Z, Li Z, Liu X, Cheng Y, Luo Y, Tong X et al (2014) Altered fecal microbiota composition associated with food allergy in infants. Appl Environ Microbiol 80(8):2546–2554

Loss G, Apprich S, Waser M, Kneifel W, Genuneit J, Büchele G et al (2011) The protective effect of farm milk consumption on childhood asthma and atopy: the GABRIELA study. J Allergy Clin Immunol 128(4):766–73.e4

Lyons A, O'Mahony D, O'Brien F, MacSharry J, Sheil B, Ceddia M et al (2010) Bacterial strain-specific induction of Foxp3+ T regulatory cells is protective in murine allergy models. Clin Exp Allergy 40(5):811–819

MacLean E, Madsen N, Vliagoftis H, Field C, Cameron L (2013) n-3 Fatty acids inhibit transcription of human IL-13: implications for development of T helper type 2 immune responses. Br J Nutr 109(6):990–1000

Macpherson AJ, Harris NL (2004) Interactions between commensal intestinal bacteria and the immune system. Nat Rev Immunol 4(6):478–485

Magne F, Abély M, Boyer F, Morville P, Pochart P, Suau A (2006) Low species diversity and high interindividual variability in faeces of preterm infants as revealed by sequences of 16S rRNA genes and PCR-temporal temperature gradient gel electrophoresis profiles. FEMS Microbiol Ecol 57(1):128–138

Maslova E, Hansen S, Strøm M, Halldorsson TI, Olsen SF (2014) Maternal intake of vitamins A, E and K in pregnancy and child allergic disease: a longitudinal study from the Danish National Birth Cohort. Br J Nutr 111(6):1096–1108

Maslowski KM, Vieira AT, Ng A, Kranich J, Sierro F, Di Y et al (2009) Regulation of inflammatory responses by gut microbiota and chemoattractant receptor GPR43. Nature 461(7268): 1282–1286

Maynard CL, Elson CO, Hatton RD, Weaver CT (2012) Reciprocal interactions of the intestinal microbiota and immune system. Nature 489(7415):231–241

Mazmanian SK, Liu CH, Tzianabos AO, Kasper DL (2005) An immunomodulatory molecule of symbiotic bacteria directs maturation of the host immune system. Cell 122(1):107–118

Menezes JS, Mucida DS, Cara DC, Alvarez-Leite JI, Russo M, Vaz NM et al (2003) Stimulation by food proteins plays a critical role in the maturation of the immune system. Int Immunol 15(3): 447–455

Miliku K, Robertson B, Sharma AK, Subbarao P, Becker AB, Mandhane PJ et al (2018) Human milk oligosaccharide profiles and food sensitization among infants in the CHILD study. Allergy 73(10):2070–2073

Moles L, Gómez M, Heilig H, Bustos G, Fuentes S, de Vos W et al (2013) Bacterial diversity in meconium of preterm neonates and evolution of their fecal microbiota during the first month of life. PLoS One 8(6):e66986

Mörbe UM, Jørgensen PB, Fenton TM, von Burg N, Riis LB, Spencer J et al (2021) Human gut-associated lymphoid tissues (GALT); diversity, structure, and function. Mucosal Immunol 14(4):793–802

Moriki D, Francino MP, Koumpagioti D, Boutopoulou B, Rufián-Henares J, Priftis KN et al (2022) The role of the gut microbiome in cow's milk allergy: a clinical approach. Nutrients 14(21):4537

Netting MJ, Middleton PF, Makrides M (2014) Does maternal diet during pregnancy and lactation affect outcomes in offspring? A systematic review of food-based approaches. Nutrition 30(11–12):1225–1241

Ni YF, Wang J, Yan XL, Tian F, Zhao JB, Wang YJ et al (2010) Histone deacetylase inhibitor, butyrate, attenuates lipopolysaccharide-induced acute lung injury in mice. Respir Res 11(1):33

Noh G, Lee JH (2011) Regulatory B cells and allergic diseases. Allergy, Asthma Immunol Res 3(3): 168

Novak N, Kraft S, Bieber T (2003) Unraveling the mission of FcepsilonRI on antigen-presenting cells. J Allergy Clin Immunol 111(1):38–44

Noval Rivas M, Burton OT, Wise P, Zhang YQ, Hobson SA, Garcia Lloret M et al (2013) A microbiota signature associated with experimental food allergy promotes allergic sensitization and anaphylaxis. J Allergy Clin Immunol 131(1):201–212

Palmer C, Bik EM, DiGiulio DB, Relman DA, Brown PO (2007) Development of the human infant intestinal microbiota. PLoS Biol 5(7):e177

Palmer DJ, Prescott SL, Perkin MR (2017) Early introduction of food reduces food allergy – pro and con. Pediatr Allergy Immunol 28(3):214–221

Perez PF, Doré JL, Leclerc M, Levenez F, Benyacoub J, Serrant P et al (2007) Bacterial imprinting of the neonatal immune system: lessons from maternal cells? Pediatrics 119(3):e724–ee32

Pulendran B, Artis D (2012) New paradigms in type 2 immunity. Science 337(6093):431–435

Quintana FJ, Basso AS, Iglesias AH, Korn T, Farez MF, Bettelli E et al (2008) Control of Treg and TH17 cell differentiation by the aryl hydrocarbon receptor. Nature 453(7191):65–71

Rey-Mariño A, Francino MP (2022) Nutrition, gut microbiota, and allergy development in infants. Nutrients 14(20):4316

Riedler J, Braun-Fahrländer C, Eder W, Schreuer M, Waser M, Maisch S et al (2001) Exposure to farming in early life and development of asthma and allergy: a cross-sectional survey. Lancet 358(9288):1129–1133

Robertson RC, Manges AR, Finlay BB, Prendergast AJ (2019) The human microbiome and child growth – first 1000 days and beyond. Trends Microbiol 27(2):131–147

Rodriguez B, Prioult G, Bibiloni R, Nicolis I, Mercenier A, Butel MJ et al (2011) Germ-free status and altered caecal subdominant microbiota are associated with a high susceptibility to cow's milk allergy in mice. FEMS Microbiol Ecol 76(1):133–144

Roduit C, Frei R, Depner M, Schaub B, Loss G, Genuneit J et al (2014) Increased food diversity in the first year of life is inversely associated with allergic diseases. J Allergy Clin Immunol 133(4):1056–1064

Roduit C, Frei R, Ferstl R, Loeliger S, Westermann P, Rhyner C et al (2019) High levels of butyrate and propionate in early life are associated with protection against atopy. Allergy 74(4):799–809

Rooks MG, Garrett WS (2016) Gut microbiota, metabolites and host immunity. Nat Rev Immunol 16(6):341–352

Röytiö H, Mokkala K, Vahlberg T, Laitinen K (2017) Dietary intake of fat and fibre according to reference values relates to higher gut microbiota richness in overweight pregnant women. Br J Nutr 118(5):343–352

Säemann MD, Böhmig GA, Osterreicher CH, Burtscher H, Parolini O, Diakos C et al (2000) Anti-inflammatory effects of sodium butyrate on human monocytes: potent inhibition of IL-12 and up-regulation of IL-10 production. FASEB J 14(15):2380–2382

Savage JH, Lee-Sarwar KA, Sordillo J, Bunyavanich S, Zhou Y, O'Connor G et al (2018) A prospective microbiome-wide association study of food sensitization and food allergy in early childhood. Allergy 73(1):145–152

Schulz VJ, van Roest M, Bol-Schoenmakers M, van Duursen MB, van den Berg M, Pieters RH et al (2013) Aryl hydrocarbon receptor activation affects the dendritic cell phenotype and function during allergic sensitization. Immunobiology 218(8):1055–1062

Segain JP, Raingeard de la Blétière D, Bourreille A, Leray V, Gervois N, Rosales C et al (2000) Butyrate inhibits inflammatory responses through NFkappaB inhibition: implications for Crohn's disease. Gut 47(3):397–403

Sekerel BE, Bingol G, Cullu Cokugras F, Cokugras H, Kansu A, Ozen H et al (2021) An expert panel statement on the beneficial effects of human milk oligosaccharides (HMOs) in early life and potential utility of HMO-supplemented infant formula in cow's milk protein allergy. J Asthma Allergy 14:1147–1164

Seppo AE, Autran CA, Bode L, Järvinen KM (2017) Human milk oligosaccharides and development of cow's milk allergy in infants. J Allergy Clin Immunol 139(2):708–11.e5

Shao Y, Forster SC, Tsaliki E, Vervier K, Strang A, Simpson N et al (2019) Stunted microbiota and opportunistic pathogen colonization in caesarean-section birth. Nature 574(7776):117–121

Sibilano R, Frossi B, Calvaruso M, Danelli L, Betto E, Dall'Agnese A et al (2012) The aryl hydrocarbon receptor modulates acute and late mast cell responses. J Immunol 189(1):120–127

Singh N, Gurav A, Sivaprakasam S, Brady E, Padia R, Shi H et al (2014) Activation of Gpr109a, receptor for niacin and the commensal metabolite butyrate, suppresses colonic inflammation and carcinogenesis. Immunity 40(1):128–139

Sjögren YM, Jenmalm MC, Böttcher MF, Björkstén B, Sverremark-Ekström E (2009) Altered early infant gut microbiota in children developing allergy up to 5 years of age. Clin Exp Allergy 39(4):518–526

Smeekens JM, Johnson-Weaver BT, Hinton AL, Azcarate-Peril MA, Moran TP, Immormino RM et al (2020) Fecal IgA, antigen absorption, and gut microbiome composition are associated with food antigen sensitization in genetically susceptible mice. Front Immunol 11:599637

Smith PM, Howitt MR, Panikov N, Michaud M, Gallini CA, Bohlooly YM et al (2013) The microbial metabolites, short-chain fatty acids, regulate colonic Treg cell homeostasis. Science 341(6145):569–573

Spergel JM, Paller AS (2003) Atopic dermatitis and the atopic march. J Allergy Clin Immunol 112(6 Suppl):S118–S127

Sprenger N, Odenwald H, Kukkonen AK, Kuitunen M, Savilahti E, Kunz C (2017) FUT2-dependent breast milk oligosaccharides and allergy at 2 and 5 years of age in infants with high hereditary allergy risk. Eur J Nutr 56(3):1293–1301

Stefka AT, Feehley T, Tripathi P, Qiu J, McCoy K, Mazmanian SK et al (2014) Commensal bacteria protect against food allergen sensitization. Proc Natl Acad Sci USA 111(36): 13145–13150

Stoodley I, Garg M, Scott H, Macdonald-Wicks L, Berthon B, Wood L (2019) Higher Omega-3 index is associated with better asthma control and lower medication dose: a cross-sectional study. Nutrients 12(1):74

Tanaka S, Kobayashi T, Songjinda P, Tateyama A, Tsubouchi M, Kiyohara C et al (2009) Influence of antibiotic exposure in the early postnatal period on the development of intestinal microbiota. FEMS Immunol Med Microbiol 56(1):80–87

Thorburn AN, McKenzie CI, Shen S, Stanley D, Macia L, Mason LJ et al (2015) Evidence that asthma is a developmental origin disease influenced by maternal diet and bacterial metabolites. Nat Commun 6:7320

Trompette A, Gollwitzer ES, Yadava K, Sichelstiel AK, Sprenger N, Ngom-Bru C et al (2014) Gut microbiota metabolism of dietary fiber influences allergic airway disease and hematopoiesis. Nat Med 20(2):159–166

Turfkruyer M, Verhasselt V (2015) Breast milk and its impact on maturation of the neonatal immune system. Curr Opin Infect Dis 28(3):199–206

Vallès Y, Gosalbes MJ, de Vries LE, Abellán JJ, Francino MP (2012) Metagenomics and development of the gut microbiota in infants. Clin Microbiol Infect 18(Suppl 4):21–26

Vallès Y, Artacho A, Pascual-García A, Ferrús ML, Gosalbes MJ, Abellán JJ et al (2014) Microbial succession in the gut: directional trends of taxonomic and functional change in a birth cohort of Spanish infants. PLoS Genet 10(6):e1004406

van den Elsen LW, Nusse Y, Balvers M, Redegeld FA, Knol EF, Garssen J et al (2013) n-3 long-chain PUFA reduce allergy-related mediator release by human mast cells in vitro via inhibition of reactive oxygen species. Br J Nutr 109(10):1821–1831

van Odijk J, Kull I, Borres MP, Brandtzaeg P, Edberg U, Hanson LA et al (2003) Breastfeeding and allergic disease: a multidisciplinary review of the literature (1966–2001) on the mode of early feeding in infancy and its impact on later atopic manifestations. Allergy 58(9):833–843

Vidal K, Donnet-Hughes A (2008) CD14: a soluble pattern recognition receptor in milk. Adv Exp Med Biol 606:195–216

Vuillermin PJ, O'Hely M, Collier F, Allen KJ, Tang MLK, Harrison LC et al (2020) Maternal carriage of Prevotella during pregnancy associates with protection against food allergy in the offspring. Nat Commun 11(1):1452

Wang S, Charbonnier L-M, Noval Rivas M, Georgiev P, Li N, Gerber G et al (2015) MyD88 adaptor-dependent microbial sensing by regulatory T cells promotes mucosal tolerance and enforces commensalism. Immunity 43(2):289–303

Weise C, Hilt K, Milovanovic M, Ernst D, Rühl R, Worm M (2011) Inhibition of IgE production by docosahexaenoic acid is mediated by direct interference with STAT6 and NFκB pathway in human B cells. J Nutr Biochem 22(3):269–275

Weise C, Ernst D, van Tol EA, Worm M (2013) Dietary polyunsaturated fatty acids and non-digestible oligosaccharides reduce dermatitis in mice. Pediatr Allergy Immunol 24(4): 361–367

Weiss ST, Mirzakhani H, Carey VJ, O'Connor GT, Zeiger RS, Bacharier LB et al (2024) Prenatal vitamin D supplementation to prevent childhood asthma: 15-year results from the Vitamin D Antenatal Asthma Reduction Trial (VDAART). J Allergy Clin Immunol 153(2):378–388

Wesemann DR, Nagler CR (2016) The microbiome, timing, and barrier function in the context of allergic disease. Immunity 44(4):728–738

Zhang X, Han Y, Tian Q, Du L, Chen L, Zhang Y et al (2023) The association between n-3 polyunsaturated fatty acid intakes and asthma in US children and adolescents: a cross-sectional study from NHANES. Pediatr Allergy Immunol 34(9):e14024

Zheng D, Liwinski T, Elinav E (2020) Interaction between microbiota and immunity in health and disease. Cell Res 30(6):492–506

Metabolic Modeling of the Human Gut Microbiota for Personalized Nutrition

6

Josefina Arcagni, Francisco J. Planes, and Telmo Blasco

Abstract

The understanding of the interaction between the human gut microbiota and diet remains one major challenge in biomedicine. In particular, the human gut microbiota degrades nutrients derived from the diet and releases key bioactive compounds that exert a strong influence on host health. Due to the high interindividual variation of the microbial composition in the human gut, personalized dietary recommendations are mandatory to efficiently promote human health. To address this complex challenge, holistic approaches that integrate experimental and computational tools have been developed in the last years. In this context, genome-scale metabolic models (GEMs) and constraint-based modeling (CBM) approaches are promising tools to widely explore the metabolic capabilities of gut microbial communities and develop personalized nutrition applications. In this chapter, we first present a brief background of GEMs and CBM, including their main concepts, assumptions, and basic mathematical models. Then, we introduce key databases and computational platforms to generate gut microbial community GEMs (co-GEMs), as well as the most relevant mathematical methods to study and explore them, including steady-state and dynamic approaches. Finally, we review seminal works that highlight the relevance of co-GEMs and CBM

J. Arcagni · T. Blasco (✉)
Tecnun School of Engineering, University of Navarra, San Sebastián, Spain
e-mail: tblasco@unav.es

F. J. Planes
Tecnun School of Engineering, University of Navarra, San Sebastián, Spain

Biomedical Engineering Center, University of Navarra, Pamplona, Navarra, Spain

Instituto de Ciencia de los Datos e Inteligencia Artificial (DATAI), University of Navarra, Pamplona, Spain

M. P. Francino, J. Á. Rufián Henares (eds.), *The Microbiome and Personalized Nutrition*, The Microbiomes of Humans, Animals, Plants, and the Environment 6,
https://doi.org/10.1007/978-3-031-93902-0_6

methods towards advancing personalized nutrition, delineating key challenges to be addressed in the coming years.

Keywords

Genome-scale metabolic models (GEMs) · Constraint-based modelling (CBM) · Gut microbiome · Personalized nutrition · Nutrient metabolism

6.1 Introduction

The human gut microbiota refers to the community of microorganisms—including bacteria, archaea, fungi, and viruses—that inhabit the human gut (Lederberg and McCray 2001; Marchesi and Ravel 2015; Lozupone et al. 2012). This complex ecosystem plays a relevant role in key physiological functions, including human metabolism, protection against pathogens, and regulating the immune response (Clemente et al. 2012; Hartstra et al. 2016). The composition and function of the human gut microbiota are altered by several factors, such as age, genetics, environment, or disease—but primarily by diet (Rinninella et al. 2023). As evidenced in a growing body of literature, the nutritional composition of foods has a direct impact on the abundance of gut microbial species and their metabolic contribution to the host's physiology. Consequently, a complete characterization of the interaction map between diet and the human gut microbiota remains a major challenge in biomedicine, with the potential to pave the way for more effective and personalized nutrition interventions.

In the context of personalized nutrition, different approaches have been applied to elucidate the complex interactions between diet and the human gut microbiota, including both experimental and computational methods (Gentile and Weir 2018; Blasco et al. 2021; Guizar-Heredia et al. 2023). In this chapter, we review systems biology approaches driven by genome-scale metabolic models (GEMs), which have undergone a significant development over the last decade (Magnúsdóttir et al. 2017; Heinken et al. 2023; Machado et al. 2018). In particular, we focus on constraint-based modeling (CBM) (van der Ark et al. 2017; Bauer and Thiele 2018; Sen and Orešič 2019), a promising framework to simulate metabolic processes in the human gut microbiota based on GEMs and increasingly available meta-omics data—mainly 16S rRNA gene sequencing and shotgun metagenomic data. We discuss current advances and challenges for CBM to become a key computational tool in the field of personalized nutrition.

For the sake of clarity, for readers unfamiliar with GEMs and CBM, we first present a brief background, including the main concepts, assumptions, and basic mathematical models. Then, we describe existing GEMs of the human gut microbiota, highlight the most relevant static and dynamic CBM tools for simulating bacterial communities in the human gut, and, finally, explore their applications in the field of health and nutrition.

6.2 Introduction to Genome-Scale Metabolic Models

Genome-scale metabolic models (GEMs) are comprehensive computational representations of the metabolic processes occurring within an organism. They capture the entire repertoire of metabolic reactions present in an organism, including the consumption and production of metabolites, transport of molecules across cellular membranes, and the regulation of metabolic pathways. Recent advancements in the fields of systems biology and omics technologies have contributed to the generation of GEMs of different cell types, including microorganisms, plants, and mammals (Henry et al. 2010; Mendoza et al. 2019; Zhang and Hua 2016). The number of GEMs has substantially increased during the last decades, starting from small reconstructions of microbial metabolism (Edwards and Palsson 1999, 2000) to more complex models that simulate human metabolic processes (Brunk et al. 2018; Robinson et al. 2020).

GEMs define a high-quality network of reactions in stoichiometric, mass, and energy balance. They are mathematically represented with the stoichiometry matrix S of dimensions $m \times n$, where m is the number of metabolites and n represents the number of reactions (Clarke 1981). Each column of matrix S contains the stoichiometric coefficients of a specific reaction in the GEM, where negative and positive values refer to reactants and products, respectively. Figure 6.1a shows an example metabolic network involving seven metabolites and nine reactions and its associated stoichiometric matrix.

High-quality GEMs typically assign reactions to the different cellular compartments and include transport reactions, referred to as exchange reactions, which enable the exchange of metabolites between compartments. Importantly, the growth medium conditions can be comprehensively integrated through these exchange reactions. In our example network in Fig. 6.1a, we have two compartments: internal and external. Specifically, we have five internal reactions (R_2, R_3, R_4, R_6, and R_7 in Fig. 6.1a) and three transport reactions (R_1, R_5, and R_8 in Fig. 6.1a). In addition, GEMs usually incorporate an artificial reaction to simulate cellular proliferation, the so-called biomass reaction (R_{BIO} in Fig. 6.1a). This reaction describes the requirements of each organism for cell growth, including the consumption of energy and particular metabolites, such as proteins, lipids, or energy, among others. Finally, GEMs provide the gene-protein-reaction (GPR) rules, which define the Boolean associations between genes, proteins and reactions within a specific organism (Feist and Palsson 2008; Gu et al. 2019; Zhang and Hua 2016). In Fig. 6.1a, we have four genes (g_1, g_2, g_3, and g_4), and R_2, for example, can be catalyzed by the proteins associated with g_1 or g_2. GPR rules allow GEMs to integrate transcriptomic and proteomic data and assess their behavior in different functional, environmental, and genetic context.

The reconstruction of GEMs starts by annotating the reference genome of a cell type, predicting the metabolic functions and retrieving the set of metabolites, reactions, and genes of the network. Typically, GEMs are manually curated at this point. Then, the draft GEM is represented through S to systematically assess its associated metabolic functions and conduct gap filling if necessary. Finally, GEMs

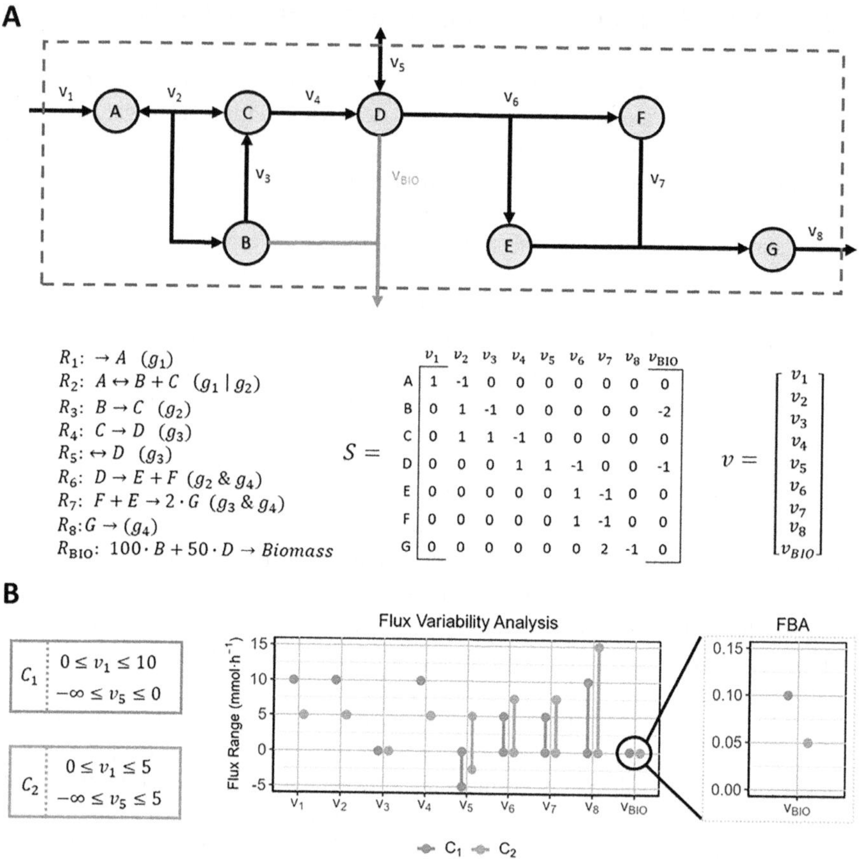

Fig. 6.1 Genome-scale metabolic models and constrained-based modeling. (**a**) Example genome-scale metabolic model. Nodes correspond to metabolites (A, B, C, D, E, F, and G) and arrows to reactions (R_1, R_2, R_3, R_4, R_5, R_6, R_7, R_8, and R_{BIO}). The reaction in green represents the biomass reaction (R_{BIO}). From the set of reactions that define the example GEM in (**a**), including their GPR rules, the stoichiometric matrix *S* and the flux vector are shown. (**b**) Illustration of flux balance analysis and flux variability analysis ($\alpha = 1$). Given two culture media C_1 and C_2 (shown in red and blue coloring, respectively), which limit the uptake flux of metabolites A and D, the minimum and maximum flux of each reaction is displayed. The flux through v_{BIO} is the result of the flux balance analysis. Abbreviations: *FBA* flux balance analysis

are validated using experimental data (Mendoza et al. 2019). The resulting high-quality GEMs can be used to analyze the production of specific metabolites, identify essential genes and reactions, and explore the metabolic capabilities of an organism in different contexts (Feist and Palsson 2008; Gu et al. 2019; Mendoza et al. 2019). These insights can be obtained thanks to the impressive developments in the area of constraint-based modeling (CBM), a relevant modeling framework for the reconstruction and analysis of GEMs. We summarize below the main concepts and mathematical foundation of CBM.

6.3 Mathematical Foundations of Constraint-Based Modeling

Constraint-based modeling (CBM) integrates an extensive array of computational methods for the reconstruction and analysis of GEMs. The basic components of CBM are the stoichiometric matrix S and reaction fluxes, mathematically represented with the flux vector v of size $n \times 1$ (Fig. 6.1a), which represents the velocity at which substrates are converted into products. Different CBM methods share the use of biochemical and biophysical constraints to interrogate the activity of GEMs across various contexts. Specifically, most CBM approaches assume two biophysical constraints: mass balance and thermodynamic constraints (Bordbar et al. 2014; Reed and Palsson 2003), which are described below.

The mass balance equation establishes that the time-dependent variation of the concentration of metabolites, integrated in the vector x of size $m \times 1$, is equal to the net balance of their production and consumption reaction fluxes:

$$\frac{dx}{dt} = S \cdot v \tag{6.1}$$

Equation (6.1) defines a system of ordinary differential equations (ODEs) that enables the study of dynamic properties of GEMs. However, CBM approaches typically assume a steady-state condition, which forces the concentration of metabolites within the model to be constant over time, Eq. (6.2). Thus, the consumption rate of every metabolite must be equal to its production rate:

$$S \cdot v = 0 \tag{6.2}$$

Moreover, CBM imposes thermodynamic constraints that limit the flux of the reactions in a specific direction. In particular, we have irreversible reactions, which are restricted to have nonnegative fluxes, and reversible reactions, which can freely take both negative and positive fluxes. In our example network in Fig. 6.1a, we have two reversible reactions (R_2 and R_5), and the rest are irreversible. Thermodynamics constraints are represented with two vectors of dimension nx1, lb and ub, that define the lower and upper bound for each reaction, respectively.

Importantly, lb and ub can also integrate restrictions about the availability of nutrients in the environment. Figure 6.1b shows two different environmental contexts that obviously lead to a different metabolic behavior. Thermodynamic and nutrient availability constraints are integrated into a set of linear inequations:

$$lb \leq v \leq ub \tag{6.3}$$

Based on Eqs. 6.2 and 6.3, a large family of mathematical methods, mainly based on optimization techniques, has been developed to study the metabolic capabilities of GEMs. One of the most popular approaches is flux balance analysis (FBA). Based on linear programming (LP), FBA searches for the flux distribution that maximizes flux through the biomass reaction, also known as the growth rate (Bordbar et al. 2014; Orth et al. 2010):

$$\textbf{LP}_1: \text{ FBA formulation}$$

$$\max z = v_{BIO} \tag{6.4}$$

$$s.t.$$

$$S \cdot v = 0 \tag{6.5}$$

$$lb \leq v \leq ub \tag{6.6}$$

Figure 6.1b shows the maximum growth rate for the two different environmental contexts considered. In the first case, when only nutrient A is available in the environment with a maximum input flux of 10 mmol/h, the maximum growth rate (v_{BIO}) is 0.1 gDW/h. In the second case, when both A and D are presented in the environment with a maximum input flux of 5 mmol/h, the maximum growth rate is 0.05 gDW/h.

A related approach is flux variability analysis (FVA) (Mahadevan and Schilling 2003). FVA directly deals with alternative optimal solutions obtained from FBA by calculating the minimum and maximum fluxes for each reaction in the network. It has a control parameter α that defines the percentage of the maximum growth rate $\left(v_{BIO}^{*}\right)$ to be satisfied. α ranges between 0 and 1, and it guarantees maximum growth rate when $\alpha = 1$. The formulation of FVA is the following:

$$\textbf{LP}_2: \text{ FVA formulation}$$

$$min/\,max_{\forall i} v_i \tag{6.7}$$

$$s.t.$$

$$S \cdot v = 0 \tag{6.8}$$

$$lb \leq v \leq ub \tag{6.9}$$

$$v_{BIO} \geq \alpha v_{BIO}^{*} \tag{6.10}$$

The application of both FBA and FVA can be extended to explore additional phenotypes beyond growth rate. Figure 6.1b shows the result of applying FVA to our two different environmental contexts. It can be observed that the second condition can reach a higher maximum production of metabolite G (15 mmol/h vs 10 mmol/h), but a lower growth rate (0.05 gDW/h vs 0.1 gDW/h). Overall, FBA and FVA are very useful approaches to explore the metabolic capabilities of GEMs.

6.4 Community Genome-Scale Metabolic Models (Co-GEMs)

GEMs have been widely applied to investigate the metabolic capabilities of single organisms under different conditions. However, the simulation of the whole metabolic activity of the gut microbial consortium is substantially more complex, requiring the integration of distinct bacterial GEMs into multispecies metabolic networks, also called community genome-scale metabolic models (co-GEMs). Co-GEMs can be generated following two different strategies: mixed-bag and compartmentalized approaches.

For illustration, Fig. 6.2a shows an example bacterial community involving two GEMs. Each GEM includes a specific subset of metabolites and reactions: The first GEM involves 7 metabolites and 10 reactions, while the second GEM involves

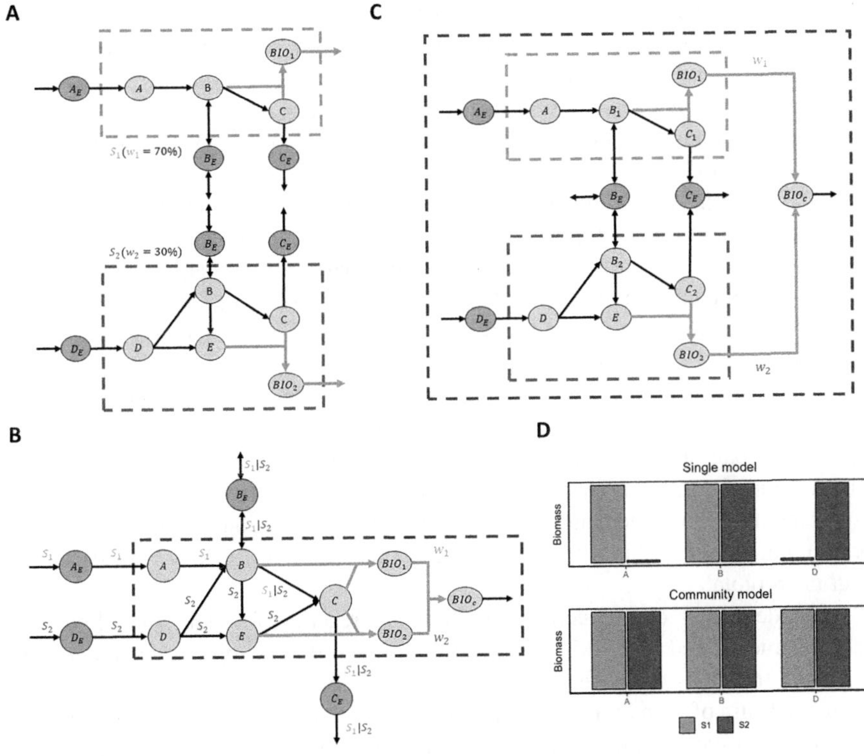

Fig. 6.2 Representation of microbial community models. (**a**) Example GEMs of two different microbes, S_1 and S_2, and their relative abundance in the consortium, w_1 and w_2, respectively. (**b**) Illustration of the resulting co-GEM according to the mixed-bag strategy. The taxonomical annotation of each reaction is displayed. (**c**) Illustration of the resulting co-GEM according to the compartmentalized strategy. The microbes within the consortium can exchange metabolites between them, as represented with extracellular metabolites $\mathbf{B_E}$ and $\mathbf{C_E}$. (**d**) The growth of each of the bacterial species if only one nutrient is present in the culture medium, when considering isolated GEMs and co-GEMs, respectively

8 metabolites and 12 reactions. We have metabolites and reactions specific to one of the GEMs, e.g., metabolite D that is present only in the second GEM, and others that are shared in the both GEMs, e.g., metabolite B. In addition, we include an artificial metabolite to model the biomass production in each GEM, denoted BIO. Finally, we assume that each GEM has two compartments: internal and external.

As shown in Fig. 6.2b, mixed-bag community models are built by merging the metabolic reactions and compounds of the GEMs involved in the microbial consortium, obtaining one internal compartment and an external compartment for the entire community (Henry et al. 2016). The boundaries between the different organisms are deleted, and the final reconstruction can be considered as a pangenome-scale metabolic model, which can be treated as a single GEM. Moreover, based on the biomass reactions of each organism, a new synthetic biomass reaction is generated to study the growth of the whole community known as the community biomass reaction. Note here that the taxonomic assignation of the microbial species involved in each metabolic reaction can be stored for further analyses (Fig. 6.2b). This strategy allows us to consider a high number of organisms in co-GEMs, and it extremely reduces the computation cost of simulations. In addition, mixed-bag community models have shown to be an adequate approach to measure output microbial metabolites in a bacterial community and to predict bacterial growth under different conditions (Greenblum et al. 2013).

In contrast, the compartmentalized approach maintains the organismal boundaries, and thus, it generates one internal compartment for each of the organisms involved in co-GEMs. Each compartment contains the metabolic reactions annotated to its associated organism. In brief, the compartmentalized approach is a concatenation of every single GEM, but, instead, a common external compartment is generated, which enables microbes to share metabolites between them and with the environment (Fig. 6.2c). As in mixed-bag models, a novel biomass reaction is also created. In addition to the different analyses covered by the mixed-bag strategy, the compartmentalized approach allows us to study the interaction of organisms by means of exchange reactions. However, this strategy is substantially more demanding in terms of computational requirements for communities with a high number of organisms, as typically found in the human gut microbiota.

The definition of the community biomass reaction for both mixed-bag and compartmentalized co-GEMs mainly follows two different approaches in the literature. In the first approach, the community biomass reaction is defined as the pondered sum of each biomass reaction by the relative abundance of their respective microbial species (Baldini et al. 2019). This reaction can be represented as:

$$\sum_i w_i \cdot BIO_i \rightarrow BIO_c \tag{6.11}$$

where w_i denotes the relative abundance of organism i, BIO_i is the artificial biomass metabolite associated with organism i, and BIO_C is the artificial community biomass reaction. This approach assumes a stable microbial community composition and balanced growth over time, which are mandatory for all the different organisms in the microbial consortium. Figure 6.2b shows this strategy to represent the community biomass reaction in the mixed-bag community model.

In addition, we can consider the community biomass as the independent contribution of the growth of the different species within the microbial community (Diener et al. 2020). According to this definition, each organism contributes to the community biomass based on its individual biomass, expressed as a percentage of the total community growth, which is determined by the relative abundance of each organism. In contrast to the previous approach, we independently define one artificial reaction for each organism (Eq. (6.12)), and thus, we do not force the growth of each organism in the bacterial consortium. Fig. 6.2c shows this strategy in the compartmentalized community model.

$$BIO_i \rightarrow w_i \cdot BIO_C \quad \forall i \tag{6.12}$$

Figure 6.2D shows the behavior of single GEMs and co-GEMs in three environmental contexts. It can be observed that the second microbe (S_2) cannot grow in isolation if only A is available in the growth medium; however, this becomes possible when S_1 and S_2 are present in the consortium. Indeed, metabolite B is produced from A by S_1 and exchanged between microbes and used to support growth in S_2. The opposite occurs when only metabolite D is available in the growth medium. In the case of metabolite B, there are no differences between single GEMs and co-GEMs. It is important to note that compartmentalized co-GEMs more naturally capture these differences than mixed-bag co-GEMs via exchange fluxes between compartments.

6.5 Co-GEMs of the Human Gut Microbiota

With the growing interest in the human gut microbiota over the last decade, substantial efforts have been dedicated to generate high-quality co-GEMs. These advances have been made possible thanks to the improvements in existing metabolic databases and the development of automatic GEM reconstruction tools. On the one hand, metabolic databases provide essential information about reactions and their functional annotation. On the other hand, automatic reconstruction tools integrate metabolic databases with functional genome annotation data to derive draft metabolic models. We describe below the most relevant databases and computational tools for the development of co-GEMs of the human gut microbiota.

One of the most popular databases is that of Model SEED (Henry et al. 2010), a foundational resource in metabolic modeling, notable for its extensive and curated library of reactions, compounds, and metabolic pathways. It was developed to facilitate the high-throughput annotation and reconstruction of single-species GEMs. To that end, Model SEED provides the functional annotation of reactions and the mapping of genes to reactions. The continually updated reaction library reflects known metabolic functions across diverse microbial taxa, establishing a robust basis for building co-GEMs. Additionally, Model SEED introduces automated gap-filling algorithms to correct missing annotations across the different metabolic pathways and maintain the network's consistency.

Recently, Faria et al. 2023 released a new version of this database, Model SEED v2, which enhances these capabilities by incorporating machine learning classifiers that provide specific biological information in the reconstruction process. Additionally, core metabolic pathways are more accurately described, and different high-quality reaction checks in the definition of reactions are implemented. These updates make Model Seed v2 an advanced tool for creating highly accurate GEMs across a wide range of microbial taxa.

KBase (Arkin et al. 2018) is a collaborative platform that extends the functionalities of Model SEED for building, curating, and analyzing GEMs, supporting both experimental and computational biology workflows. KBase integrates information from Model SEED, UniProt (Bateman 2019), and KEGG (Kanehisa 2000; Kanehisa et al. 2016) databases to automatically generate GEMs, enhancing both the accuracy and the scope of the resulting GEMs. In addition, KBase integrates microbial genome databases and provides a wide variety of functionalities for genome and metagenome assembly and annotation.

Another large effort in the literature is Assembly of Gut Organisms Through Reconstruction and Analysis (AGORA), which is a comprehensive collection of manually curated GEMs of the human gut microbiota (Magnúsdóttir et al. 2017). Initially, AGORA contained 773 gut microbe GEMs, which later expanded to include 818 species. However, the recent publication of AGORA2 has further increased this number to 7302 GEMs of organisms present in the human gut microbiota (Heinken et al. 2023).

AGORA models are generated using a semi-automated pipeline, where draft metabolic reconstructions are initially created from the Model SEED and KBase databases. Then, comprehensive quality controls and data-driven refinement of key metabolic pathways are performed over these draft reconstructions in order to obtain the resulting GEMs (Thiele and Palsson 2010). Consequently, the final GEMs exhibit high taxonomic and metabolic diversity, with detailed annotations at gene, metabolite, and reaction levels. The diversity of AGORA models allows a wide range of analyses across numerous human gut microbiota species and supports comprehensive studies of metabolic interactions within complex microbial communities in various clinical conditions. For instance, AGORA models have been applied to predict dietary supplements for Crohn's disease patients using metagenomic data (Bauer and Thiele 2018).

Another outstanding approach is Constraint-Based Reconstruction and Analysis of the Virtual Microbial Ecosystem (CarveMe) (Machado et al. 2018), a comprehensive repository of GEMs, specifically focused on microbial communities, including 5587 reconstructions. CarveMe presents an automated reconstruction pipeline of GEMs that is fundamentally different from the tools discussed above. In particular, CarveMe integrates high-quality GEMs available in the BiGG database (Schellenberger et al. 2010) to build a universal metabolic model. Then, according to the genomic evidence of reactions, the universal model is carved by removing reactions unlikely to occur in the organism under consideration. Importantly, CarveMe enables a faster generation of high-quality GEMs and co-GEMs. It has already been applied to study the overgrowth of *Candida* species and prevent systemic candidiasis in lung cancer (Seelbinder et al. 2023).

Using AGORA as a reference point, AGORA-based REconstruction for Diet Analysis (AGREDA) (Blasco et al. 2021) enhances the understanding of gut microbiota metabolism in relation to input dietary compounds. AGREDA successfully integrates degradation pathways for 209 diet-derived metabolites, including 179 phenolic compounds, which are strongly related to human health and nutrition. This expansion significantly contributes to the study of the interaction between diet and gut microbiota, as previous GEMs in AGORA included a limited number of dietary compounds compared to comprehensive databases of personalized nutrition.

In order to build AGREDA models, we presented a reconstruction algorithm that follows a mixed-bag strategy and directly extracts co-GEMs that are able to degrade key input dietary compounds. Specifically, a gap-filling algorithm is applied to integrate input dietary compounds with AGORA by including reactions from Model SEED and literature on gut microbiota metabolism. In addition, the AGREDA pipeline undergoes rigorous validation to ensure mass and stoichiometric balance. As a result, AGREDA outperforms existing reconstructions in predicting dietary impacts on gut microbiota and serves as a foundation for future research in personalized nutrition and gut health.

Overall, as summarized in Fig. 6.3a, these combined efforts provide a comprehensive framework for developing high-quality co-GEMs in the human gut microbiota. Their integration and continuous refinement enhance the quality and functional relevance of co-GEMs, enabling the exploration of the intricate metabolic networks of microbial ecosystems in the human gut microbiota.

6.6 Constraint-Based Modeling of Co-GEMs

Constraint-based modeling of co-GEMs introduces an additional layer of complexity with respect to single GEMs. As shown in Fig. 6.2, co-GEMs consider the growth of different organisms in the consortium and their metabolic complementation via exchange metabolites, which requires specific CBM strategies for flux prediction. Specifically, CBM approaches to co-GEMs can be classified into two classes: steady-state and dynamic. The main difference between these strategies lies in their treatment of spatiotemporal dynamics. Steady-state approaches explore the metabolic capabilities of organisms at a given time point, whereas dynamic methods investigate the evolution of the microbial ecosystem through time and/or space. Accordingly, each method is better suited to address different biological questions. We review below key CBM approaches for both the strategies.

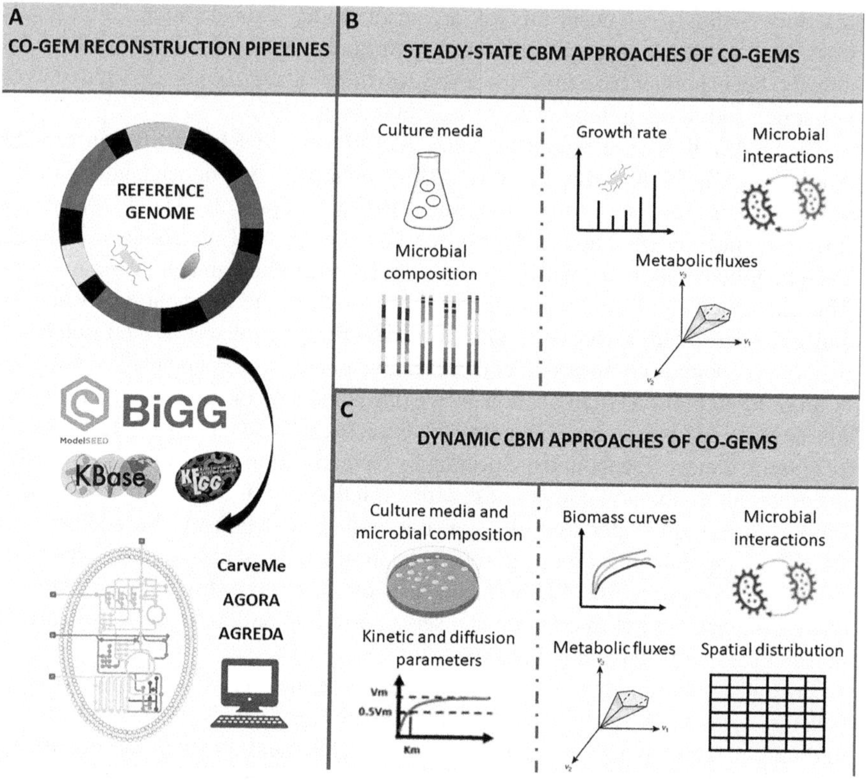

Fig. 6.3 Summary of co-GEMs of the human gut microbiota and CBM approaches. (**a**) Simplified scheme of the co-GEM reconstruction pipelines of the human gut microbiota, including the most relevant metabolic databases (Model SEED, KEGG, Kbase, and BiGG) and published co-GEM models (AGORA, CarveMe, and AGREDA). (**b**) Input and output data involved in steady-state CBM approaches to co-GEMs. (**c**) Input and output data involved in dynamic CBM approaches to co-GEMs

6.6.1 Steady-State CBM Approaches to Co-GEMs

This strategy does not consider time evolution, and, thus, standard CBM methods, such as FBA and FVA, can be directly applied to co-GEMs. The primary distinction among steady-state approaches lies in their modeling of the growth of the different organisms within the community, as well as the mathematical framework to predict reaction fluxes, including linear, nonlinear, and multilevel optimization techniques. Table 6.1 summarizes the main features of different steady-state CBM approaches to co-GEMs.

Figure 6.3b illustrates the different input and output data involved in steady-state CBM approaches to co-GEMs. These techniques commonly require microbial relative abundance data and the culture media composition. The integration of this information with co-GEMs leads to context-specific microbial co-GEMs, which can

Table 6.1 Summary of the most relevant CBM approaches to co-GEMs

Method	Year	Type	Description
cFBA	2013	Steady-state	Nonlinear optimization problem that maximizes the community growth rate with unknown microbial composition. It guarantees stable microbial composition and balanced growth.
SteadyCom	2017	Steady-state	Nonconvex optimization problem that maximizes the community growth rate with unknown microbial composition.
MMT	2019	Steady-state	Linear optimization problem that maximizes the community growth rate. It guarantees stable microbial composition and balanced growth.
OptCom	2012	Steady-state	Multi-level optimization problem that maximizes independent taxa biomass flux in the inner layer, and then, community biomass flux in the outer layer.
MICOM	2020	Steady-state	Regularized optimization problem, where the maximum biomass flux of the community and minimum 2-norm of individual biomass fluxes are traded-off.
DMMM	2011	Temporal	The first multispecies dynamic FBA method. It uses explicit Euler method to solve differential equations.
μBialSim	2020	Temporal	Multispecies dynamic FBA method that uses an augmented Euler method to avoid negative concentrations.
ORKA	2020	Temporal	Multispecies dynamic FBA method that uses the Runge-Kutta method to avoid negative concentrations.
DFBALab	2018	Temporal	Multispecies dynamic FBA method that considers a second optimization term in FBA to address the issue of multiple optimal solutions.
d-OptCom	2014	Temporal	Extension of OptCom that considers time derivatives of the concentration of biomass and extracellular metabolites in the outer problem.
COMETS	2014	Spatiotemporal	Combination of dFBA and biophysical models that simulate the diffusion of biomass and metabolites.
BacArena	2017	Spatiotemporal	Combination of dFBA and agent-based modelling, including population dynamics events, such as cellular movement, duplication, apoptosis and cross-feeding interactions.
BN-BacArena	2024	Spatiotemporal	Extension of BacArena that includes microbe-microbe and nutrient-microbe regulatory interactions obtained from a data-driven Bayesian Network framework.

CBM approaches are divided into steady-state (blue) and dynamic (orange). We included four columns: (1) name of the method; (2) year of publication; (3) classification into steady-state or dynamic approaches, and, in this last case, whether (or not) spatial dependence is included in the method (temporal vs. spatiotemporal); and (4) description of the approach, including the type of mathematical optimization model

be used to study the growth rate of the different organisms of the microbial consortium, updates in the microbial composition, and microbe–microbe pairwise interactions, among others.

In brief, some approaches assume the same growth rate for each organism in the community, called the community growth rate (μ), leading to a stable microbial composition and balanced growth:

$$v_{bio}^{i} = w_i \cdot \mu_i, \quad \mu_1 = \mu_2 = \ldots = \mu \tag{6.13}$$

This strategy was generally developed in two early approaches: community flux balance analysis (cFBA) and SteadyCom (Chan et al. 2017). In both the cases, the community growth rate μ is maximized, but the microbial composition w is assumed unknown, leading to nonlinear and nonconvex complex optimization problems, respectively. More recently, the Microbiome Modelling Toolbox (MMT) (Baldini et al. 2019), which makes use of the community biomass reaction defined in Eq. (6.11), simplified this nonlinear optimization problem into linear programming by assuming the microbial composition known from meta-omics data. This approach has been widely applied to different questions, e.g., analysis of the perturbed gut microbiota in a cohort of gouty arthritis patients (Henson 2021).

Other approaches do not force the growth of different organisms in the community at equal rate, and, thus, a stable microbial composition is not guaranteed. Here, the community objective is to maximize the biomass production:

$$\max \sum_{i=i}^{n} v_{bio}^{i} \tag{6.14}$$

This community objective was first used in OptCom (Zomorrodi and Maranas 2012), which defined a multilevel optimization model, where the inner problems maximize the biomass of each organism and the outer problem maximizes the community biomass production. However, this approach is not computationally tractable for the complex communities found in the gut microbiota. To overcome this issue, MiCOM (Diener et al. 2020) proposes a two-norm regularization strategy for biomass fluxes (Eq. 6.15) while allowing a suboptimal community growth (Eq. 6.16) using the parameter α:

$$min \sum_{i=i}^{n} {v_{bio}^{i}}^{2} \tag{6.15}$$

$$\sum_{i=i}^{n} v_{bio}^{i} \geq \alpha \cdot v_{bio}^{*}, \quad v_{bio}^{*} = \max \sum_{i=i}^{n} v_{bio}^{i} \tag{6.16}$$

The two-norm minimization in MiCOM directly deals with the large underdetermination observed in optimal flux distributions in previous approaches, i.e., different flux distributions that reach the same objective value. Additionally, it avoids solutions with unrealistic growth of a limited number of highly abundant taxa and the zero growth of many others (Colarusso et al. 2021). MiCOM has been applied to study different biological questions, including the production of short-chain fatty acids (SCFA) under different dietary interventions (Quinn-Bohmann et al. 2023).

6.6.2 Dynamic CBM Approaches to Co-GEMs

Dynamical modeling of gut microbial communities enables the study of metabolic fluxes through time and/or space. An influential CBM approach is dynamic flux balance analysis (dFBA) (Mahadevan et al. 2002), which assumes steady state for intracellular metabolites but a time-dependent variation of extracellular metabolites. Specifically, reaction fluxes are simulated via FBA, and time-dependent availability of nutrients and biomass growth are determined by solving their associated ordinary differential equations (ODEs). Dynamic CBM approaches to co-GEMs differ in the numerical method selected to solve the underlying differential equations, optimization framework to calculate reaction fluxes, and the consideration of spatial variations and events. Table 6.1 summarizes the most relevant dynamic CBM approaches to co-GEMs.

As shown in Fig. 6.3c, dynamic CBM approaches to co-GEMs typically require the culture media and microbial composition as input data. However, in contrast to steady-state CBM approaches, dynamic algorithms can also integrate kinetic parameters and diffusion coefficients of the metabolites within the culture media in order to more accurately predict reaction fluxes in co-GEMs. As a result, dynamic CBM approaches enable the simulation of biomass curves, microbial interactions, and the spatial distribution of the microorganisms and nutrients, among others.

This family of models was introduced in Zhuang et al. 2011, where the first dynamic multispecies metabolic modeling (DMMM) approach was presented. DMMM extends dFBA at the community level. Specifically, DMMM independently applies FBA to each individual taxon in the community at every time point. By integrating the resulting growth rates of all microbes within the community, as well as the production/consumption rates of the metabolites in the environment, the DMMM framework updates the extracellular metabolites and biomass concentrations using an explicit Euler method.

The main limitation of the DMMM framework is that extracellular metabolites could reach negative concentrations over time (Colarusso et al. 2021). This problem can be solved by using smaller time intervals, but it extremely increases computation time. To deal with this issue, μBialSim (Popp and Centler 2020) and ORKA (Schroeder and Saha 2020) implement different numerical methods to limit the time step reduction for intervals where the predicted metabolite concentrations would be negative (see Table 6.1).

As in steady-state approaches, multiple optimal flux vectors can be obtained from the application of FBA to each individual taxon, and the selection of any of them significantly influences time-dependent concentrations of extracellular metabolites. To deal with this issue, DFBALab (Gomez and Barton 2018) minimizes the flux of user-selected exchange reactions shared among different members of the microbial consortia, in addition to the primary objective of maximizing their growth rate. In addition, d-OptCom (Zomorrodi et al. 2014) extends OptCom to consider time derivative concentrations of biomass and extracellular metabolites in the outer problem, including kinetic information for specific exchange reactions.

Another relevant family of dynamic CBM tools includes those that consider spatial constraints, referred here as spatiotemporal approaches. They extend dynamic FBA methods to better understand how nutrient availability and mass diffusion affect microbial community dynamics in a spatially structured environment.

For illustration, Fig. 6.4 shows a toy example of spatiotemporal approaches. In essence, these methods define a two-dimensional space, typically divided into grid

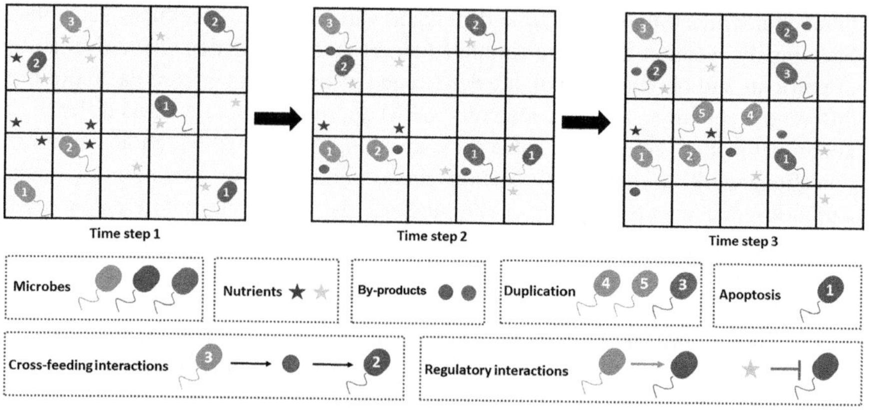

Fig. 6.4 Toy example of CBM spatiotemporal approaches to co-GEMs. 5 × 5 grid involving three microbes (microbe icons: green, blue, and red), two nutrients (star icons: purple and yellow), and two by-products (dot icons: blue and red). The evolution of a microbial ecosystem is illustrated during three-time steps. Different events can be considered, including nutrient consumption and production, cell duplication and apoptosis, cross-feeding interactions, and microbe–microbe and nutrient–microbe regulatory interactions

cells, where different microbes and nutrients are distributed. In our example, three different microbes are considered (microbe icons: green, blue, and red), which are initially fed by two nutrients (star icons: yellow and purple). The metabolic behavior of each microbe is modeled through its associated GEM. At each time step, the growth rate of each organism is determined using FBA, and the nutrient concentration in the environment is updated based on the FBA-derived flux distribution. Depending on the accumulated biomass of each organism, the composition of the microbial community evolves over time. As shown in Fig. 6.4, after three iterations, three new microbes are generated (two for microbe green and one for microbe red) and one microbe (microbe blue) undergoes cellular death due to the culture medium composition in its particular environment.

In this context, COMETS (Harcombe et al. 2014; Dukovski et al. 2021) constitutes the first spatiotemporal approach for the simulation of microbial communities. COMETS combines dynamic multispecies FBA with molecular and cellular biophysical models. Specifically, COMETS studies the evolution of the microbial consortia and the accumulated biomass in each grid cell, which may change depending on the FBA result. Additionally, nutrients can be exchanged with the neighboring grid cells via diffusion.

In the same direction, BacArena (Bauer et al. 2017) is another relevant approach. BacArena combines dFBA methods and agent-based modeling. By considering microbes as individual agents, BacArena is able to model various population dynamic events that extend the capabilities of COMETS, including cellular movement, duplication, apoptosis, and cross-feeding interactions, among others. Recently, Blasco et al. 2024 developed BN-BacArena, an extension of BacArena that accounts for microbe–microbe and nutrient–microbe interactions by means of a Bayesian network, which iteratively corrects the growth rate of each organism and improves the predicted microbial relative abundance profiles over time.

Cross-feeding interactions can be captured as the number of nutrient exchanges between microbial species. As shown in Fig. 6.4, the green microbe exchanges the blue by-product that is metabolized by the blue microbe in order to generate biomass. In addition, the growth rate of the different microorganisms can be affected by other agents. For example, the yellow nutrient inhibits the growth of the blue microbe, while it grows faster in the presence of the green microbe (Fig. 6.4).

6.7 Personalized Nutrition Applications of Co-GEMs

Recent research underscores the necessity of personalized dietary approaches that consider human gut microbiota. For instance, Wastyk et al. 2021 examined how plant-based fiber and fermented foods influence the human microbiome and immune system in healthy adults, demonstrating that specific dietary interventions can effectively modulate immune status. Furthermore, Zeevi et al. 2015 highlighted the variability in glycemic responses among individuals by studying glucose levels in a large cohort. Their findings revealed significant differences in responses to identical meals, suggesting that universal dietary recommendations may have limited utility. These studies collectively illustrate the importance of tailoring dietary interventions to individual needs to optimize health outcomes.

Human gut microbiota co-GEMs are promising in the development of personalized dietary recommendations tailored to individual gut microbiota compositions. By simulating metabolic interactions within the gut environment, co-GEMs facilitate a comprehensive understanding of how different dietary inputs affect both microbial and host responses. This capability has fostered promising research that links dietary habits with microbial metabolism and health outcomes, revealing insights beyond standard nutritional guidelines. We summarize below some relevant works in this direction.

In an early study presented in Heinken et al. 2013, co-GEMs were used to model human and mouse gut microbiomes under high-protein, high-fat, and ketogenic diets, revealing shifts in microbial interactions and nutrient metabolism. Specifically, gut microbial co-GEMs were constructed from Model SEED and BiGG databases. Then, cross-feeding interactions and metabolic exchanges were analyzed with FVA. This work highlights the capacity of co-GEMs to comprehensively illustrate diet–host–microbiome dynamics, with potential implications on designing dietary interventions to promote gut health.

Later, Henson and Phalak 2017, applied co-GEMs to investigate the dysbiosis associated with inflammatory bowel disease (IBD). Their study concentrated on how the depletion of *Faecalibacterium prausnitzii* and the proliferation of *Escherichia coli* can exacerbate IBD symptoms, particularly under various nutrient conditions, including diets with different carbohydrate and protein concentrations. This work emphasized the impact of dietary composition on disease progression. By conducting precise simulations of dietary effects with DFBALab, co-GEMs presented a valuable approach for understanding how nutrient adjustments may promote a healthier microbial balance and potentially reduce or delay IBD symptoms.

Similarly, Baldini et al. 2020 demonstrated the potential of co-GEMs to link microbial dysbiosis to Parkinson's disease (PD). Specifically, they identified alterations in the microbiome that correlate with PD. By modeling the gut microbiota composition with MMT and making comparisons between PD and control patients, this work highlights the potential of co-GEMs to guide dietary adjustments that could mitigate PD-related symptoms, such as constipation, by leveraging individual gut microbial profiles.

Other works have made use of co-GEMs and CBM approaches to link human gut microbiota and drug metabolism (Pant et al. 2023; Heinken et al. 2023). Particularly, Rosario et al. 2018, hypothesized that microbial dysbiosis was associated with metformin treatment, a standard treatment for type 2 diabetes. The authors observed that patients under metformin treatment exhibit abnormal abundance of several representative bacterial strains, whose metabolic by-products exerted a strong influence in type 2 diabetes development. Then, AGORA-based co-GEMs and MMT were applied to assess the metabolic interactions with the host's metabolic environment, shedding light on personalized nutrition strategies to support diabetic patients based on their microbiome.

More recently, Quinn-Bohmann et al. 2023, explored co-GEMs in the human gut microbiota to predict SCFA production in response to various dietary interventions. In particular, personalized AGORA community metabolic models were generated, and the production of acetate, butyrate, and propionate was measured across different samples through MiCOM. In vitro validation showed a significant correlation between the observed SCFA concentrations and the predicted production fluxes across several dietary interventions. This study additionally explored the effects of prebiotic and probiotic supplements as potential interventions to enhance gut microbiome health in these contexts.

Despite this tremendous effort and promising works, there are several challenges that must be addressed for developing comprehensive and accurate models of the human gut microbiota. First, there are still whole families of compounds whose degradation pathways are not included in co-GEMs, which limits their use in nutritional applications. Second, there are few databases that comprehensively describe the nutritional composition of different diets, and the ones available, such as Virtual Metabolic Human (Noronha et al. 2019), lack many key nutrients that are essential for human health and nutrition.

AGREDA constitutes the biggest step toward achieving an accurate description of degradation pathways of compounds in the human diet, allowing for diet-specific reconstructions. However, new methodologies are necessary to improve further the information within co-GEMs of the human gut microbiota. In this context, enzyme promiscuity methods, such as RetroPath (Carbonell et al. 2014) or PROXIMAL (Yousofshahi et al. 2015; Balzerani et al. 2024), aim to predict interactions between target substrates and enzymes to decipher novel reactions. We have successfully extended AGREDA by incorporating new reactions predicted by RetroPath and PRXOXIMAL involved in the degradation of phenolic compounds in the human gut microbiota (Balzerani et al. 2022). These studies open the door to the development of similar works that address other classes of compounds relevant for the gut microbiota.

Beyond the knowledge of cellular metabolism in the human gut microbiota and the definition of input compounds from the diet, it is also necessary to consider the digestive processes that food naturally undergoes before arriving to the intestine. In particular, it is critical to consider the absorption rate of each nutrient to precisely define the intestinal environment and measure the microbial by-products. Moreover, the specific combinations of metabolites that are beneficial for human health are not yet fully understood; therefore, it is still unknown what species should increase or decrease through interventions to promote human health.

Overall, we anticipate a relevant role of co-GEMs of the human gut microbiota in the development of personalized dietary recommendations, particularly within the context of the diet–microbiome–health axis. By simulating metabolic outcomes and elucidating the complex interactions between diet, microbiota, and host physiology, co-GEMs allow researchers to generate insights that are designed to improve individual metabolic health and disease prevention. As research in co-GEMs advances, these models are set to become essential tools for guiding dietary choices that not only enhance metabolic health but also support the prevention of diet-related diseases: a key step toward personalized nutrition.

References

Arkin AP et al (2018) KBase: the United States department of energy systems biology knowledgebase. Nat Biotechnol 36:566–569

Baldini F et al (2019) The microbiome modeling toolbox: from microbial interactions to personalized microbial communities. Bioinformatics 35:2332–2334

Baldini F et al (2020) Parkinson's disease-associated alterations of the gut microbiome predict disease-relevant changes in metabolic functions. BMC Biol 18:1–21

Balzerani F et al (2022) Prediction of degradation pathways of phenolic compounds in the human gut microbiota through enzyme promiscuity methods. npj Syst Biol Appl 8:24

Balzerani F et al (2024) Extending PROXIMAL to predict degradation pathways of phenolic compounds in the human gut microbiota. npj Syst Biol Appl 10:1–11

Bateman A (2019) UniProt: a worldwide hub of protein knowledge. Nucleic Acids Res 47:D506–D515

Bauer E, Thiele I (2018) From metagenomic data to personalized in silico microbiotas: predicting dietary supplements for Crohn's disease. npj Syst Biol Appl 4:1–26

Bauer E et al (2017) BacArena: individual-based metabolic modeling of heterogeneous microbes in complex communities. PLoS Comput Biol 13:1–22

Blasco T et al (2021) An extended reconstruction of human gut microbiota metabolism of dietary compounds. Nat Commun 12:1–12

Blasco T et al (2024) BN-BacArena: Bayesian network extension of BacArena for the dynamic simulation of microbial communities. Bioinformatics 40:btae266

Bordbar A et al (2014) Constraint-based models predict metabolic and associated cellular functions. Nat Rev Genet 15:107–120

Brunk E et al (2018) Recon3D: a resource enabling a three-dimensional view of gene variation in Human metabolism. Nat Biotechnol 36:272–281
Carbonell P et al (2014) Retropath: automated pipeline for embedded metabolic circuits. ACS Synth Biol 3:565–577
Chan SHJ et al (2017) SteadyCom: predicting microbial abundances while ensuring community stability. PLoS Comput Biol 13:1–25
Clarke BL (1981) Complete set of steady states for the general stoichiometric dynamical system. J Chem Phys 75:4970–4979
Clemente JC et al (2012) The impact of the gut microbiota on human health: an integrative view. Cell 148:1258–1270
Colarusso AV et al (2021) Computational modeling of metabolism in microbial communities on a genome-scale. Curr Opin Syst Biol 26:46–57
Diener C et al (2020) MICOM: metagenome-scale modeling to infer metabolic interactions in the gut microbiota. mSystems 5:e00606-19
Dukovski I et al (2021) A metabolic modeling platform for the computation of microbial ecosystems in time and space (COMETS). Nat Protoc 16:5030–5082
Edwards JS, Palsson BO (1999) Systems properties of the Haemophilus influenzae Rd. metabolic genotype. J Biol Chem 274:17410–17416
Edwards JS, Palsson BO (2000) Metabolic flux balance analysis and the in silico analysis of Escherichia coli K-12 gene deletions. BMC Bioinformatics 1:Article 1
Faria JP et al. (2023) ModelSEED v2: high-throughput genome-scale metabolic model reconstruction with enhanced energy biosynthesis pathway prediction. bioRxiv, 2023.10.04.556561
Feist AM, Palsson B (2008) The growing scope of applications of genome-scale metabolic reconstructions using Escherichia coli. Nat Biotechnol 26:659–667
Gentile CL, Weir TL (2018) The gut microbiota at the intersection of diet and human health. Science 362:776–780
Gomez JA, Barton PI (2018) Dynamic flux balance analysis using DFBAlab. Methods Mol Biol 1716:353–370
Greenblum S et al (2013) Towards a predictive systems-level model of the human microbiome: Progress, challenges, and opportunities. Curr Opin Biotechnol 24:810–820
Gu C et al (2019) Current status and applications of genome-scale metabolic models. Genome Biol 20:1–18
Guizar-Heredia R et al (2023) A new approach to personalized nutrition: postprandial glycemic response and its relationship to gut microbiota. Arch Med Res 54:176–188
Harcombe WR et al (2014) Metabolic resource allocation in individual microbes determines ecosystem interactions and spatial dynamics. Cell Rep **7**:1104–1115
Hartstra AV et al (2016) Interplay between gut microbiota, its metabolites and human metabolism: dissecting cause from consequence. Trends Food Sci Technol 57:233–243
Heinken A et al (2013) Systems-level characterization of a host-microbe metabolic symbiosis in the mammalian gut. Gut Microbes 4:28–40
Heinken A et al (2023) Genome-scale metabolic reconstruction of 7302 human microorganisms for personalized medicine. Nat Biotechnol 41:1320–1331
Henry CS et al (2010) High-throughput generation, optimization and analysis of genome-scale metabolic models. Nat Biotechnol 28:977–982
Henry CS et al (2016) Microbial community metabolic modeling: a community data-driven network reconstruction. J Cell Physiol 231:2339–2345
Henson MA (2021) Interrogation of the perturbed gut microbiota in gouty arthritis patients through in silico metabolic modeling. Eng Life Sci 21:489–501
Henson MA, Phalak P (2017) Microbiota dysbiosis in inflammatory bowel diseases: in silico investigation of the oxygen hypothesis. BMC Syst Biol 11:1–15
Kanehisa M (2000) KEGG: Kyoto Encyclopedia of Genes and Genomes. Nucleic Acids Res 28:27–30

Kanehisa M et al (2016) KEGG as a reference resource for gene and protein annotation. Nucleic Acids Res 44:D457–D462
Lederberg BJ, McCray AT (2001) 'Ome sweet' omics— a genealogical treasury of words. Science 15:8
Lozupone CA et al (2012) Diversity, stability and resilience of the human gut microbiota. Nature 489:220–230
Machado D et al (2018) Fast automated reconstruction of genome-scale metabolic models for microbial species and communities. Nucleic Acids Res 46:7542–7553
Magnúsdóttir S et al (2017) Generation of genome-scale metabolic reconstructions for 773 members of the human gut microbiota. Nat Biotechnol 35:81–89
Mahadevan R, Schilling CH (2003) The effects of alternate optimal solutions in constraint-based genome-scale metabolic models. Metab Eng 5:264–276
Mahadevan R et al (2002) Dynamic flux balance analysis of diauxic growth in Escherichia coli. Biophys J 83:1331–1340
Marchesi JR, Ravel J (2015) The vocabulary of microbiome research: a proposal. Microbiome 3:1–3
Mendoza SN et al (2019) A systematic assessment of current genome-scale metabolic reconstruction tools. Genome Biol 20:1–20
Noronha A et al (2019) The Virtual Metabolic Human database: Integrating human and gut microbiome metabolism with nutrition and disease. Nucleic Acids Res 47:D614–D624
Orth JD et al (2010) What is flux balance analysis? Nat Biotechnol 28:245–248
Pant A et al (2023) Human gut microbiota and drug metabolism. Microb Ecol 86:97–111
Popp D, Centler F (2020) μBialSim: constraint-based dynamic simulation of complex microbiomes. Front Bioeng Biotechnol 8:574
Quinn-Bohmann N et al (2023) Microbial community-scale metabolic modeling predicts personalized short chain fatty acid production profiles in the human gut. bioRxiv, 9
Reed JL, Palsson B (2003) Thirteen years of building constraint-based in silico models of Escherichia coli. J Bacteriol 185:2692–2699
Rinninella E et al (2023) The role of diet in shaping human gut microbiota. Best Pract Res Clin Gastroenterol 62–63:101828
Robinson JL et al (2020) An atlas of human metabolism. Sci Signal 13:eaaz1482
Rosario D et al (2018) Understanding the representative gut microbiota dysbiosis in metformin-treated Type 2 diabetes patients using genome-scale metabolic modeling. FrontPhysiol 9:1–14
Schellenberger J et al (2010) BiGG: a biochemical genetic and genomic knowledgebase of large scale metabolic reconstructions. BMC Bioinformatics 11:213
Schroeder WL, Saha R (2020) Introducing an optimization- and explicit Runge-Kutta- based approach to perform dynamic flux balance analysis. Sci Rep 10:1–28
Seelbinder B et al (2023) Candida expansion in the gut of lung cancer patients associates with an ecological signature that supports growth under dysbiotic conditions. Nat Commun 14:1–15
Sen P, Orešič M (2019) Metabolic modeling of human gut microbiota on a genome scale: an overview. Metabolites 9:22
Thiele I, Palsson B (2010) A protocol for generating a high-quality genome-scale metabolic reconstruction. Nat Protoc 5:93–121
van der Ark KCH et al (2017) More than just a gut feeling: constraint-based genome-scale metabolic models for predicting functions of human intestinal microbes. Microbiome 5:78
Wastyk HC et al (2021) Gut-microbiota-targeted diets modulate human immune status. Cell 184: 4137–4153.e14
Yousofshahi M et al (2015) PROXIMAL: a method for prediction of xenobiotic metabolism. BMC Syst Biol 9:1–17
Zeevi D et al (2015) Personalized nutrition by prediction of glycemic responses. Cell 163:1079–1094
Zhang C, Hua Q (2016) Applications of genome-scale metabolic models in biotechnology and systems medicine. Front Physiol 6:1–8

Zhuang K et al (2011) Genome-scale dynamic modeling of the competition between Rhodoferax and Geobacter in anoxic subsurface environments. ISME J. 5:305–316

Zomorrodi AR, Maranas CD (2012) OptCom: a multi-level optimization framework for the metabolic modeling and analysis of microbial communities. PLoS Comput Biol 8:e1002363

Zomorrodi AR et al (2014) D-OptCom: dynamic multi-level and multi-objective metabolic modeling of microbial communities. ACS Synth Biol 3:247–257

From Microbiota to Diet: Transferring Information on the Personal Microbiota to the Individual's Diet

7

Verónica González-Vigil

Abstract

The growing body of research on human gut microbiota has revealed its crucial role in regulating numerous physiological processes, including digestion, immunity, and even mental health (Kau et al. Nature 474(7351):327–336, 2011). Understanding how individual microbiota profiles interact with dietary components has opened new avenues in personalized nutrition, enabling the design of individually tailored diets to improve health outcomes and prevent diseases (Johnson et al. Cell Host Microbe 25(6):789–802.e5, 2019). This chapter explores how the microbiome, through the production of key metabolites, can be modulated by diet to positively influence health (Sonnenburg and Bäckhed, Nature 535(7610):56–64, 2016). Improvements in the available food composition tables and advancements in microbiome-sequencing technologies, such as 16S rRNA gene sequencing and metagenomics (Lloyd-Price et al. Nature 569(7758): 655–662, 2019), along with computational models, have made it possible to customize dietary recommendations, considering each individual's unique microbiome composition (Heinken et al. Nat Biotechnol 41(9):1320–1331, 2023). We will discuss how the Stance4Health project has addressed this challenge by developing the "microbiota scores" algorithm, a tool that evaluates and ranks foods according to their capacity to nourish the microbiota in a healthy manner (Hinojosa-Nogueira et al. Nutrients 15(2):276, 2023) and that is integrated into the personalized dietary recommendation system of the i-Diet S4H app. This system facilitates the creation of personalized menus that not only meet nutritional requirements but also promote a favorable and sustainable intestinal environment in the long term (Dello Russo et al. Foods 11(10):1480,

V. González-Vigil (✉)
GSN NutriTech S.L., Alicante, Spain
e-mail: vgonzalez@gsnsoft.es

M. P. Francino, J. Á. Rufián Henares (eds.), *The Microbiome and Personalized Nutrition*, The Microbiomes of Humans, Animals, Plants, and the Environment 6,
https://doi.org/10.1007/978-3-031-93902-0_7

2022). By integrating microbiota data into dietary recommendations, we can advance toward truly personalized nutrition, offering tailored interventions that enhance an individual's metabolic, immune, and neurological health (Valdes et al. BMJ:k2179, 2018).

Keywords

Gut microbiome · Personalized nutrition · Dietary intervention · Precision medicine · Diet · microbiota interaction · Dietary recommendations · Recommendation algorithms

7.1 Introduction

In recent years, the relationship between the human microbiome and health has garnered increasing interest, catalyzing a significant shift in the understanding of nutrition (Kau et al. 2011; Valdes et al. 2018). The gut microbiota (GM), a complex ecosystem harboring billions of microorganisms, has emerged as a central element in personalized medicine and nutrition. This chapter focuses on how advances in nutrition, computational sciences, biotechnology, and bioinformatics are revolutionizing the personalization of nutrition based on individual microbial profiles (Ordovas et al. 2018).

Progress in food composition tables (FCTs), whose importance has often been underestimated, has been essential to this process. The creation of more comprehensive, reliable FCTs enriched with data on nutrients and bioactive compounds has created a snowball effect, fostering the development of numerous new tools, as they form the foundation for many calculations needed to understand interactions and personalize diets (Hinojosa-Nogueira et al. 2021).

Combined with innovations in genomic sequencing, such as 16S rRNA gene analysis and metagenomics, our understanding of the gut microbiota has expanded significantly (Qin et al. 2010; iHMP Research Network Consortium 2019). These technologies enable not only the identification of species present but also the comprehension of their metabolic functions. This highlights how the microbiota reflects and influences an individual's dietary and environmental interactions, affecting nutrient metabolism (Lloyd-Price et al. 2016). The transformation of dietary fibers and polyphenols by the microbiota, for instance, produces important metabolites such as short-chain fatty acids, which have beneficial effects on metabolic health and inflammation (Deehan and Walter 2016; Sonnenburg and Bäckhed 2016). Advanced bioinformatics tools, such as AGORA and AGREDA and their evolved versions, have been essential for enhancing our ability to analyze interactions between diet and microbiota (Heinken et al. 2023). These platforms have facilitated the exploration of how specific dietary compounds can be influenced by the microbiota. Additionally, q2-metnet, a QIIME2 plug-in, allows the contextualization of 16S rRNA gene sequencing data in AGORA and AGREDA (Balzerani et al. 2024).

Simultaneously, personalized dietary recommendation systems have evolved, becoming increasingly sophisticated with the development of evolutionary algorithms, machine learning, and enhanced computational power (Zeevi et al. 2015). These systems now integrate multiple sources of information, such as individual microbiota data, nutritional requirements, dietary habits, clinical conditions, and expert knowledge, to design highly tailored and nutritionally balanced diets (Sempionatto 2021). These diets not only optimize an individual's metabolic and immune health but also do so in a sustainable manner over the long term.

In this chapter, we will explore how these technologies enable an unprecedented integration of microbiota data into clinical and research practice, paving the way for truly personalized nutrition and shaping a future in which nutrition and microbiota are intrinsically linked to improve health and prevent disease (Blasco et al. 2021; Dello Russo et al. 2022).

7.2 Understanding Key Factors

The fundamental principle of personalized nutrition is that there is no universal diet that works for everyone. The metabolic response to food varies significantly between individuals, largely due to the influence of the gut microbiome (Balzerani et al. 2022). This approach considers multiple factors, including genetic profile, food preferences, health status, and increasingly the microbiome composition. By adjusting an individual's diet based on these factors, it is possible to enhance overall health and reduce the risk of developing diseases such as obesity, type 2 diabetes, and cardiovascular diseases (Blasco et al. 2021).

One of the main objectives of incorporating individual microbiota information into personalized nutrition is to modulate the microbiome to optimize its functionality. Since diet is one of the most significant factors influencing microbiome composition, adjusting the intake of specific macronutrients and micronutrients can promote a healthier microbial community. For example, diets rich in fiber and polyphenols can increase microbial diversity and foster the production of beneficial metabolites, such as short-chain fatty acids (SCFAs) (Blasco et al. 2021).

As we gain a better understanding of how dietary components interact with the microbiome, we also learn how these interactions impact human health, from metabolism to immune and neurological functions. In recent years, many factors or elements have contributed to advancing this understanding, generating important synergies that have led to the development of specific strategies and tools for incorporating individual microbiota into personalized nutritional recommendations. These can be grouped into four main categories:

- Food composition tables or food composition databases (FCTs/FCDs)
- Sequencing technologies
- Bioinformatics and computational tools
- Recommendation systems

Often underestimated, FCTs/FCDs are the cornerstone, as their nutrient and bioactive compound data are the essential foundation for bioinformatic calculations and dietary assessments. Therefore, having complete, reliable, and suitable data sources is essential for developing any quality nutritional calculation or assessment. However, a gap exists in nutritional data, and many European databases are limited in the number of foods and/or nutrient information they provide. Additionally, they are often outdated due to the enormous time and financial resources required to update them. This is why many studies and tools rely on open-access tables from other countries (frequently from the US Food and Agriculture Organization [FAO] or the United States Department of Agriculture [USDA]), without considering factors like biodiversity, which can affect the quality of calculations. However, with technology that reduces analysis costs and computational advances, a growing number of countries are updating their FCTs/FCDs, such as the McCance and Widdowson's tables, NEVO (Netherlands), and Frida Food Data (Denmark). Notably, the European project Stance4Health developed a comprehensive database containing over 800 nutrients and bioactive compounds, allergen data, seasonality, and other traditionally unavailable, but nutritionally valuable data (Hinojosa-Nogueira et al. 2021).

Advances in next-generation sequencing (NGS) technologies, such as 16S rRNA gene sequencing and metagenomic sequencing, have revolutionized our ability to analyze the composition and function of the gut microbiota in depth (Qin et al. 2010; iHMP Research Network Consortium 2019). These techniques have not only significantly improved the efficiency and reduced the costs of analysis but also transformed our understanding of the microbial ecosystems in the human gut. They have also enabled the development of new tools to interpret the generated data, allowing substantial progress in biomedical research and the application of microbiome-based personalized treatments (Lloyd-Price et al. 2016).

The development and implementation of bioinformatics and computational tools have been fundamental in translating the complexity of microbiological data into practical, personalized nutritional recommendations. Platforms such as AGORA and AGREDA, in particular, have revolutionized this field. AGORA, a metabolic reconstruction, provides a framework for understanding how microbes within the microbiome contribute to human metabolic processes. This platform has allowed researchers to simulate and predict metabolic interactions between diet and microbiota, which is crucial for formulating dietary recommendations that aim to optimize microbiome health. AGREDA, on the other hand, extends AGORA's capabilities by integrating additional data on diet–microbiota interactions, including a broader range of metabolites and metabolic reactions. This expansion has significantly enhanced the accuracy of predictions regarding how certain foods may influence the gut microbiota. With the ability to process and analyze these complex datasets, AGREDA enables even more refined diet personalization, targeting a microbial balance that supports overall health. In both the cases, we have witnessed how advances in computational capabilities and new-generation knowledge have driven their rapid evolution, resulting in new, improved versions.

These platforms, along with other bioinformatics tools such as MetaPhlAn, Kraken 2/Bracken, HUMAnN, and QIIME, allow researchers and nutritionists to detail the composition and function of microbial communities (Bolyen et al. 2019; Manghi et al. 2023). Each of these tools has been developed and enhanced over the years to keep pace with advances in sequencing and data analysis, making microbiome research increasingly accessible and detailed, while also fostering the development of complementary plug-ins or tools, such as q2-metnet, a QIIME2 plug-in (Balzerani et al. 2024) that enables the contextualization of 16S rRNA gene sequencing data in AGORA and AGREDA. These are just some examples, although computational methods continue to emerge, advancing our understanding of cellular metabolism to guide costly and time-demanding experimentation. As previously mentioned, we are in an ecosystem where each step forward feeds the knowledge of other areas or tools, creating important synergies and promoting their application in fields such as personalized nutrition.

Integrating this vast amount of data and expert knowledge into understandable recommendations tailored to the needs and preferences of consumers is no easy task; in this context, recommendation systems have become a fundamental tool (Zeevi et al. 2015). These systems are designed to transform the complex and vast information derived from microbiota and nutrition studies into personalized recommendations, thereby facilitating the adoption of diets that promote a healthy microbial balance.

Moreover, to efficiently manage this complexity, these systems use advanced algorithms that can evolve and adapt to the changing needs of the user. A relevant example is the use of evolutionary algorithms to generate complex, structured recommendations, which have proven effective, especially in dietary recommendations that depend on multiple constraints and personal characteristics (Korem et al. 2017).

It is essential to keep in mind that, when making these recommendations, not only health-related factors but also a balance with other aspects, such as individual habits, preferences, tastes, and lifestyle, should be considered (Valdes et al. 2018). This balance is essential to achieve adherence, the key element for a real impact on health. The effectiveness of any personalized nutritional intervention depends on its ability to integrate naturally into an individual's daily life, promoting lasting changes that genuinely impact their health in the long term, thus ensuring that the effort invested in its design and development is not in vain.

7.3 Translating Microbiota Information to Dietary Recommendations: The Stance4Health Project

The overarching goal of the Stance4Health Project (Smart Technologies for personAlised Nutrition and Consumer Engagement 2018; https://www.stance4health.com/) was to develop a comprehensive Smart Personalized Nutrition (SPN) service that, through the use of mobile technologies and personalized food production, optimizes gut microbiota activity and fosters long-term consumer engagement

(Dello Russo et al. 2022). This approach combines the precision of personalized nutrition with accessible and practical tools, enabling users to adopt and maintain dietary habits that not only promote gut health but also contribute to overall well-being in a sustainable manner.

One of the key innovations of Stance4Health is the "gut microbiota score" or S4H-GM score, a fundamental component of the recommendation engine in the app. This algorithm ranks foods based on their ability to nourish and optimize the gut microbiota, allowing the creation of personalized menus that not only meet nutritional requirements but also promote a beneficial and sustainable intestinal environment over the long term (Formisano et al. 2024).

7.3.1 Calculation of the S4H-GM Score

The S4H-GM score defines the impact of food consumption on an individual's gut microbiota metabolic function, which can produce beneficial or harmful metabolites for the host's physiology. This score can be positive or negative, reflecting the healthy or unhealthy effect of ingested food micronutrients on gut microbiota metabolic function (Blasco et al. 2021).

The S4H-GM score is calculated from the S4H-GM score per ingredient, which determines the impact of 100 g of a particular ingredient on gut microbiota metabolic function. To evaluate the impact of 100 g of a specific ingredient on an individual's gut microbiota metabolic function, a genome-scale metabolic model was constructed based on:

- Genome-scale metabolic reconstruction: The score is based on AGREDA, a detailed, high-quality metabolic reconstruction of the human microbiota. This model simulates how a specific food affects the production of critical health-related metabolites, such as short-chain fatty acids (SCFAs), known for their beneficial effects on inflammation and metabolism (Heinken et al. 2023).
- 16S rRNA gene sequencing data from the individual's fecal sample. Advances in sequencing technologies, such as 16S rRNA gene sequencing, have provided critical data for personalizing nutrition based on individual microbiota profiles (Balzerani et al. 2024).
- Micronutrient composition of 100 g of a specific ingredient: Each ingredient's evaluation includes an analysis of its fiber, polyphenol, and other bioactive compound content, which are essential determinants of microbiota composition and functionality. The creation of the Stance4Health food composition database was vital for this work, as existing FCTs/FCDs were limited (Hinojosa-Nogueira et al. 2021).

An optimization algorithm was then applied to predict the net production of key metabolites for human health. Finally, the S4H-GM score for the ingredient was obtained by normalizing and averaging the production of these metabolites, considering that beneficial metabolites are given a positive sign and harmful metabolites are given a negative one. Taking into account the quantities of different ingredients consumed daily by the individual, the S4H-GM score was calculated as a weighted sum of the S4H-GM score per ingredient. This score is integrated into the app's recommendation engine.

7.3.2 Personalized Menu Recommendation Generator System

The Stance4Health menu recommendation generator system is the final step in translating microbiota and nutritional information into practical, personalized recommendations for the user. When eating habits are adapted to each person's needs and preferences, they not only foster the adoption of a healthy lifestyle but also contribute to sustainable nutrition that can be maintained over the long term. This approach is particularly relevant to the gut microbiota, whose composition and functionality depend on a consistent intake of nutrients that promote microbial diversity and the ecological balance of the gut (Sonnenburg and Bäckhed 2016). Adherence to a healthy and appropriate diet is essential for the effectiveness of any nutritional intervention. However, one of the greatest challenges in nutrition is people's ability to maintain dietary changes over time. Personalizing menus helps to overcome this obstacle by tailoring dietary recommendations to the user's specific tastes, needs, and goals, thereby increasing motivation and commitment. In summary, it is not enough to create nutritionally perfect menus for the microbiota. If we truly want to impact the user's health, it is essential that they also be appetizing and align with the user's lifestyle, habits, tastes, and preferences (Zeevi et al. 2015).

This system is capable of utilizing a variety of input sources (both data-driven and expert knowledge) and producing a complex, structured output. An evolutionary approach has been employed to combine various unit elements within a flexible structure, creating an initial set of configurable complex elements. Then, a content-based approach refines (in relation to preferences) these candidates to offer a final recommendation in the form of a complete weekly menu.

This process is developed through four main stages: data source integration, evolutionary generation, content-based refinement, and a final recommendation, accompanied by a feedback loop to optimize the system (Fig. 7.1).

In the first stage, Integration of Multiple Input Sources, the recommendation system is enriched by a diverse combination of data and multidisciplinary expert knowledge. The data sources include, first, user data, encompassing dietary preferences, dietary restrictions, and health goals. Second, nutritional and food data detail the composition of foods, their nutritional values, and the presence of bioactive compounds that could impact the user's health. Finally, the system incorporates expert knowledge in key areas, such as gut microbiota, dietetics, and general health, providing a solid foundation of reliable, up-to-date information that enhances the recommendation process.

The second stage, known as the Evolutionary Approach to Generating Complex Structures, uses evolutionary algorithms to combine unit elements, such as specific foods or food groups, into menu structures that form the basis of the initial

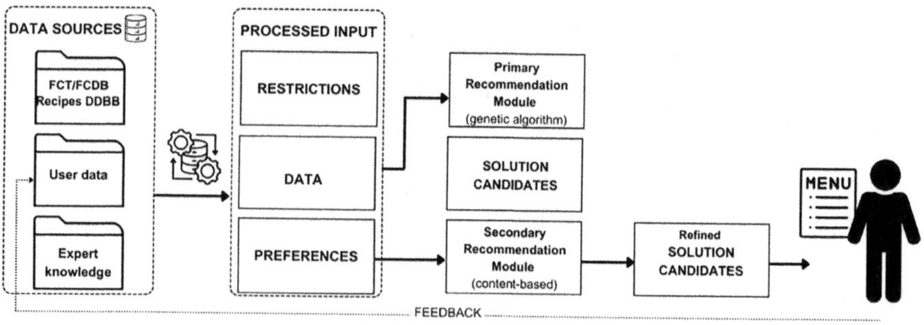

Fig. 7.1 Basic scheme of the Stance4Health recommender system

recommendation. This evolutionary approach allows the system to simulate multiple combinations and variations of foods until a balanced menu structure is achieved that respects both nutritional recommendations and the user's personal restrictions. During this phase, the algorithms evolve and optimize combinations, generating an initial set of menu candidates. This set is also highly configurable and adaptable, allowing for more precise personalization in the later stages of the process.

The third stage, Content-Based Refinement, applies a content-driven approach to adjust the initial menu options according to the user's preferences and past behaviors. This refinement considers aspects such as personal tastes, preferred textures, and meal preparation times, aligning the menus with the user's lifestyle. Through this approach, the system can prioritize the menu candidates that best match personal preferences, thus promoting greater adherence and satisfaction. The goal of this stage is to deliver a more fine-tuned recommendation aligned with the user's specific interests, significantly enhancing their experience.

Finally, the fourth stage culminates in the Final Recommendation and Feedback Loop. The system presents a personalized menu recommendation that not only meets nutritional and health requirements but also fully respects the user's preferences. The feedback loop in the S4H system is a fundamental tool that continuously refines and adapts menu recommendations based on user interactions and changes in their health metrics. This process begins with data collection during app use, where users record new recipes, tastes, preferences, or changes in their parameters and provide feedback on each recommended menu, indicating their satisfaction with the suggestions, modifications to dishes, or other adjustments. The personalized menu generator system is integrated into the i-Diet S4H mobile application (Hinojosa-Nogueira et al. 2023), providing an accessible and a user-friendly interface. The app can monitor health-related metrics over time, such as weight, pathologies, and allergies through manual input, or other health parameters such as blood pressure, lean mass, fat mass, and pulse through integration with the Toledo T9 activity wristband.

Subsequently, the system analyzes the feedback to identify consistent patterns in the user's preferences and behavior or deviations from nutritional and microbiological requirements. Based on these data, the system can adjust future recommendations and propose new suggestions that help the user regain balance. To implement these changes, the recommendation algorithms are continuously updated based on processed data, adjusting parameters such as nutritional goals, meal configuration, desired portion sizes, inclusion or exclusion of specific foods or recipes, and option variety. This dynamic personalization responds to changes in the user's lifestyle and health needs.

The adjusted data and new user preferences are reintegrated into the menu generation process, ensuring that each new set of recommendations aligns with the user's current needs. This iterative process produces menus, designed to favorably modulate the microbiota, which are increasingly personalized, resulting in continuous improvement in recommendation quality and thus greater satisfaction and adherence from the user over time.

Additionally, the recommendation system provides daily tips on nutrition, diet, or physical activity through the app, educating users on new elements of food and nutrition. It also includes alerts to show users if their behavior has impacted a set of relevant nutrients and how to correct it in the future.

This innovative approach in the Stance4Health project focuses on creating a personalized nutrition solution that not only adapts to individual nutritional needs but also improves gut health, empowering users through informed dietary recommendations and promoting sustainable change in their eating habits.

To validate and evaluate the effectiveness of the Smart Personalized Nutrition (SPN) service in modifying the gut microbiota (GM) composition and metabolic function and in empowering the consumer through the adoption of innovative technology, Stance4Health conducted a clinical trial. This trial was implemented in two contexts: normal-weight and overweight adults in Spain and children with obesity, gluten-related disorders, or cow's milk allergy/intolerance in Greece (Dello Russo et al. 2022). The intervention, centered on the use of the SPN, assessed the service's ability to personalize nutrition based on the microbiota profile and foster sustainable dietary habit change. The trial was registered under the number ISRCTN63745549, ensuring transparency and adherence to ethical and methodological standards of clinical research.

7.4 Challenges and Future Perspectives

The approach of personalizing diets based on the microbiota and the individual profile of each user has shown great potential to improve human health. However, several challenges must be addressed to optimize this technology and make it effective and accessible at both clinical and public health levels (Ordovas et al. 2018).

7.4.1 Challenges in Microbiota-Based Dietary Personalization

Dependence on Food Composition Databases Food composition databases, essential for calculating the impact of nutrients on the microbiota, have limitations in terms of the inclusion and updating of specific ingredients. Many databases lack information beyond the nutritional labels of commercial foods, which are predominantly found in household shopping baskets, impacting the accuracy of dietary recommendations. To overcome this limitation, it is necessary to improve the coverage and quality of food databases by integrating specific nutrient information that can be used in computational models evaluating the impact of diet on the microbiota.

Limitations in Standardization and Interindividual Variability Despite advances in 16S rRNA gene sequencing and metagenomics, the lack of standardization in microbiota analysis methods and the high interindividual variability remain significant barriers. The composition of the microbiota can vary widely between individuals due to factors such as genetics, environment, lifestyle, and habitual diet (Kau et al. 2011). This variability makes it challenging to generalize results and apply a uniform personalization approach, especially in large and diverse populations. Standardizing microbiota analysis methods and creating reference profiles that account for these variations are essential to improve the precision of microbiota-based dietary recommendations (Lloyd-Price et al. 2019).

Complexity in Metabolite and Biomarker Analysis The production of metabolites by the microbiota is a complex process that varies depending on the foods consumed and each individual's microbial composition (Sonnenburg and Bäckhed 2016). Identifying and quantifying key metabolites accurately and consistently remain a challenge, especially in a clinical setting where rapid and reliable methods are required. Furthermore, the ability to assess the impact of these metabolites on overall health and microbiota function requires more refined and accessible tools and predictive models to understand how dietary interventions affect the microbiota over time. This creates a significant demand for computational resources and the need to integrate multidisciplinary knowledge from the fields such as microbiology, genetics, and informatics.

Adaptation of Real-Time Recommendations The gut microbiota is dynamic and can change in response to diet over relatively short periods (Johnson et al. 2019). Adapting nutrition recommendations to respond in real time to these changes requires an agile analysis and feedback system that can quickly capture alterations in microbiota composition and adjust dietary recommendations almost immediately. Currently, achieving real-time adaptation is complex, both technically and logistically, as well as economically.

Complexity and High Development Cost of SPNs The technical complexity and high associated costs of developing SPNs are challenges. These systems require multidisciplinary teams that include experts in microbiology, bioinformatics, artificial intelligence (AI), nutrition, and social sciences, among others (Hinojosa-Nogueira et al. 2023). Integrating complex microbiological data into dietary recommendations demands a constant and costly collaboration, especially to ensure the scientific accuracy and relevance of the system. Additionally, the process involves high costs in clinical trials and data infrastructure, necessary to ensure the precision and effectiveness of the recommendations, adding a significant layer of technical and financial challenges.

User Education and Long-Term Adherence User education and empowerment are essential for personalized nutrition services to be effective (Sempionatto et al. 2021). Long-term adherence is one of the most critical challenges, as the success of any nutritional intervention largely depends on users' understanding and valuing the benefits of their dietary decisions (Valdes et al. 2018). It is essential to communicate the concepts and health benefits effectively to the population.

7.4.2 Opportunities and Future Perspectives

The field of microbiota-based personalized nutrition is in a phase of rapid growth, offering multiple opportunities to improve health through precise, tailored dietary interventions. As some of the current challenges are overcome, new opportunities and perspectives emerge that will allow these advances to be consolidated into clinical practice, democratize access, and maximize their impact on global health.

Expansion of Knowledge in Nutrition and Microbiota The growth of microbiota and food composition databases presents an unprecedented opportunity for expanding scientific knowledge in this field. These databases, continuously enriched with new information on bioactive compounds and microbiota, allow for a deeper understanding of how specific dietary components affect gut and overall health (Hinojosa-Nogueira et al. 2021). Furthermore, international collaboration to standardize data and facilitate information-sharing across global databases can accelerate advancements in personalized nutrition. This expansion of knowledge is crucial to developing highly accurate algorithms and predictive models that support scientifically and clinically valuable dietary recommendations.

Integration of Artificial Intelligence and Machine Learning Artificial intelligence (AI) and machine learning offer significant opportunities to enhance the precision and efficiency of personalized dietary recommendations. By analyzing large volumes of microbiota and dietary habit data, AI models can identify complex patterns that may be difficult to detect with conventional methods (Zeevi et al. 2015). In the future, these models will be able to create more accurate and adapted dietary recommendations, facilitating the design of diets that maximize the benefits for each individual's microbiota.

Incorporation of New Biomarkers and Multi-Omics Data Incorporating additional biomarkers, such as transcriptomic, proteomic, and metabolomic data, into microbiota evaluation would allow for a more holistic understanding of the interaction between diet and gut health. The integration of multi-omics data will provide a more detailed view of microbial functions and enable more specific recommendations based on the physiological and metabolic responses of the user to certain foods. This multidimensional approach could be key to advancing the personalization and effectiveness of dietary interventions (Hasin et al. 2017).

Development of Noninvasive Tools for Real-Time Monitoring The implementation of noninvasive technologies for continuous microbiome and health marker monitoring represents a promising perspective. With diagnostic tools that enable real-time microbiota data collection, such as breath or saliva metabolite sensors, it would be possible to adjust recommendations instantaneously, optimizing the user's dietary response. These tools would allow for even more precise personalization of nutrition and enable users to monitor their progress in a more immediate, visual way, which is crucial for increasing commitment and adherence to recommendations (Sempionatto et al. 2021).

Integration into Clinical Practice and Chronic Disease Prevention Incorporating personalized nutrition into clinical practice offers key opportunities to improve the prevention and treatment of chronic diseases such as obesity, type 2 diabetes, and cardiovascular diseases. By tailoring dietary recommendations to the unique characteristics of each individual's microbiota and genetics, health professionals can offer preventive interventions that optimize metabolic health and reduce the risk of chronic illnesses (Valdes et al. 2018). This approach also has significant implications for managing specific disorders, such as inflammatory bowel disease or food allergies, where microbiota modulation can play a crucial role in symptom improvement and patient quality of life (Cani and de Vos 2017).

Democratization of Access to Personalized Nutrition As more accessible and low-cost technologies for microbiota sequencing and data analysis are developed, microbiota-based personalized nutrition could become available to a larger number of people. This would allow populations from diverse socioeconomic and geographic backgrounds to access personalized dietary recommendations that promote optimal gut health, helping reduce health disparities and promoting greater equity in access to quality care (Khazen et al. 2020).

User Empowerment and Education Personalized nutrition services offer a valuable opportunity to empower users in managing their health (Sonnenburg and Bäckhed 2016). By providing educational tools and interactive content, these services can help users understand the relationship between their diet, microbiota, and overall well-being. This empowerment fosters greater adherence and long-term commitment, allowing users to make informed decisions and actively participate in maintaining their health. Additionally, access to continuous reports and feedback through the app reinforces understanding of how their dietary choices impact their microbiota and overall health, promoting lasting habit changes and a preventive health approach (Dello Russo 2022).

As science and technology progress, microbiota-based personalized nutrition is set to become an essential tool for public health and preventive medicine. The possibility of integrating these services into clinical settings, schools, and communities, combined with the growth of accessible diagnostic tools and real-time monitoring technologies, promises a future where health interventions can be continuously tailored to each individual's unique needs (Ordovas et al. 2018).

Future perspectives include the potential to develop algorithms that consider not only the microbiota and diet but also other health data or environmental factors (Korem et al. 2017). This comprehensive approach will enable even more complete and accurate health recommendations, elevating personalized nutrition to a broader and more effective level of precision medicine.

In summary, the opportunities in personalized nutrition are vast, and the future perspectives show a path of continuous evolution. By integrating innovation, scientific collaboration, and accessibility, personalized nutrition services have the potential to transform how we approach health and well-being, empowering individuals to live healthier, more fulfilling lives in harmony with their unique biology.

7.5 Conclusion

This chapter has explored how microbiome-based personalized nutrition is emerging as a promising field in precision medicine, with the potential to revolutionize our understanding and approach to individual health. The interaction between diet and the microbiome has proven to be key to understanding how foods influence microbial composition and, consequently, various health aspects, such as metabolism, the immune system, and neurological function (Kau et al. 2011; Sonnenburg and Bäckhed 2016). Through diet personalization, based on individual microbiological profiles, specific dietary interventions can be formulated to promote the production of beneficial metabolites and reduce the risk of chronic diseases (Valdes et al. 2018).

Despite the rapid and significant advancements in recent years, this field still faces substantial challenges, including interindividual variability, lack of standardization in analysis methods, and difficulty integrating these advances into daily clinical practice.

Nevertheless, the future of nutritional personalization is promising. Ongoing advancements in sequencing technologies and the use of more precise predictive models based on artificial intelligence will enable a deeper understanding of the interaction between the microbiome and diet, facilitating the creation of highly personalized dietary recommendations (Zeevi et al. 2015; Korem et al. 2017). Over time, microbiome-based personalized nutrition is likely to become a common tool in precision medicine, helping to improve overall health and prevent a range of chronic diseases (Ordovas et al. 2018).

Integrating microbiome information into precision medicine has the potential to radically change the approach to disease prevention and treatment, providing health professionals with new tools to offer more specific and effective treatments (Cani and de Vos 2017). Ultimately, the use of microbiome data in nutrition personalization has the power not only to optimize health at the individual level but also to transform how we approach healthcare as a whole (Dello Russo et al. 2022).

References

Balzerani F, Hinojosa-Nogueira D, Cendoya X, Blasco T, Pérez-Burillo S, Apaolaza I, Francino MP, Rufián-Henares JÁ, Planes FJ (2022) Prediction of degradation pathways of phenolic compounds in the human gut microbiota through enzyme promiscuity methods. Npj Sys Biol Appl 8(1):1–9. https://doi.org/10.1038/s41540-022-00234-9

Balzerani F, Blasco T, Pérez-Burillo S, Francino MP, Rufián-Henares JÁ, Valcarcel L, Planes FJ (2024) q2-metnet: QIIME2 package to analyze 16S rRNA data via high-quality metabolic reconstructions of the human gut microbiota. Bioinformatics. (Oxford, England):btae455. https://doi.org/10.1093/bioinformatics/btae455

Blasco T, Pérez-Burillo S, Balzerani F, Hinojosa-Nogueira D, Lerma-Aguilera A, Pastoriza S, Cendoya X, Rubio Á, Gosalbes MJ, Jiménez-Hernández N, Francino MP, Apaolaza I, Rufián-Henares JÁ, Planes FJ (2021) An extended reconstruction of human gut microbiota metabolism of dietary compounds. Nat Commun 12(1):1–12. https://doi.org/10.1038/s41467-021-25056-x

Bolyen E, Rideout JR, Dillon MR, Bokulich NA, Abnet CC, Al-Ghalith GA, Alexander H, Alm EJ, Arumugam M, Asnicar F, Bai Y, Bisanz JE, Bittinger K, Brejnrod A, Brislawn CJ, Brown CT, Callahan BJ, Caraballo-Rodríguez AM, Chase J et al (2019) Reproducible, interactive, scalable and extensible microbiome data science using QIIME 2. Nat Biotechnol 37(8):852–857. https://doi.org/10.1038/s41587-019-0209-9

Cani PD, de Vos WM (2017) Next-generation beneficial microbes: the case of Akkermansia muciniphila. Front Microbiol 8. https://doi.org/10.3389/fmicb.2017.01765

Deehan EC, Walter J (2016) The fiber gap and the disappearing gut microbiome: implications for human nutrition. Trends Endocrinol Metab 27(5):239–242. https://doi.org/10.1016/j.tem.2016.03.001

Dello Russo M, Russo P, Rufián-Henares JÁ, Hinojosa-Nogueira D, Pérez-Burillo S, de la Cueva SP, Rohn S, Fatouros A, Douros K, González-Vigil V, Epstein D, Francino MP, Siani A, Lauria F (2022) The Stance4Health project: evaluating a smart personalised nutrition service for gut microbiota modulation in normal- and overweight adults and children with obesity, gluten-related disorders or allergy/intolerance to cow's milk. Foods (Basel, Switzerland) 11(10):1480. https://doi.org/10.3390/foods11101480

Formisano A, Russo MD, Russo P, Siani A, Hinojosa-Nogueira D, Navajas-Porras B, Toledano-Marín Á, Pastoriza S, Blasco T, Lerma-Aguilera A, Francino MP, Planes FJ, González-Vigil V, Rufián-Henares JÁ, Lauria F (2024) Development and validation of a self-administered semi-quantitative food frequency questionnaire focused on gut microbiota: the Stance4Health-FFQ. Nutrients 16(23):4064. https://doi.org/10.3390/nu16234064

Hasin Y, Seldin M, Lusis A (2017) Multi-omics approaches to disease. Genome Biol 18(1):83. https://doi.org/10.1186/s13059-017-1215-1

Heinken A, Hertel J, Acharya G, Ravcheev DA, Nyga M, Okpala OE, Hogan M, Magnúsdóttir S, Martinelli F, Nap B, Preciat G, Edirisinghe JN, Henry CS, Fleming RMT, Thiele I (2023) Genome-scale metabolic reconstruction of 7,302 human microorganisms for personalized medicine. Nat Biotechnol 41(9):1320–1331. https://doi.org/10.1038/s41587-022-01628-0

Hinojosa-Nogueira D, Pérez-Burillo S, Navajas-Porras B, Ortiz-Viso B, de la Cueva SP, Lauria F, Fatouros A, Priftis KN, González-Vigil V, Rufián-Henares JÁ (2021) Development of an unified food composition database for the European project "Stance4Health". Nutrients 13(12):4206. https://doi.org/10.3390/nu13124206

Hinojosa-Nogueira D, Ortiz-Viso B, Navajas-Porras B, Pérez-Burillo S, González-Vigil V, de la Cueva SP, Rufián-Henares JÁ (2023) Stance4Health nutritional APP: a path to personalized smart nutrition. Nutrients 15(2):276. https://doi.org/10.3390/nu15020276

Johnson AJ, Vangay P, Al-Ghalith GA, Hillmann BM, Ward TL, Shields-Cutler RR, Kim AD, Shmagel AK, Syed AN, Walter J, Menon R, Koecher K, Knights D (2019) Daily sampling reveals personalized diet-microbiome associations in humans. Cell Host Microbe 25(6): 789–802.e5. https://doi.org/10.1016/j.chom.2019.05.005

Kau AL, Ahern PP, Griffin NW, Goodman AL, Gordon JI (2011) Human nutrition, the gut microbiome and the immune system. Nature 474(7351):327–336. https://doi.org/10.1038/nature10213

Khazen W, Jeanne J-F, Demaretz L, Schäfer F, Fagherazzi G (2020) Rethinking the use of mobile apps for dietary assessment in medical research. J Med Internet Res 22(6):e15619. https://doi.org/10.2196/15619

Korem T, Zeevi D, Zmora N, Weissbrod O, Bar N, Lotan-Pompan M, Avnit-Sagi T, Kosower N, Malka G, Rein M, Suez J, Goldberg BZ, Weinberger A, Levy AA, Elinav E, Segal E (2017) Bread affects clinical parameters and induces gut microbiome-associated personal glycemic responses. Cell Metab 25(6):1243–1253.e5. https://doi.org/10.1016/j.cmet.2017.05.002

Lloyd-Price J, Abu-Ali G, Huttenhower C (2016) The healthy human microbiome. Genome Med 8(1):51. https://doi.org/10.1186/s13073-016-0307-y

Lloyd-Price J, Arze C, Ananthakrishnan AN, Schirmer M, Avila-Pacheco J, Poon TW, Andrews E, Ajami NJ, Bonham KS, Brislawn CJ, Casero D, Courtney H, Gonzalez A, Graeber TG, Hall AB, Lake K, Landers CJ, Mallick H, Plichta DR et al (2019) Multi-omics of the gut microbial ecosystem in inflammatory bowel diseases. Nature 569(7758):655–662. https://doi.org/10.1038/s41586-019-1237-9

Manghi P, Blanco-Míguez A, Manara S, NabiNejad A, Cumbo F, Beghini F, Armanini F, Golzato D, Huang KD, Thomas AM, Piccinno G, Punčochář M, Zolfo M, Lesker TR, Bredon M, Planchais J, Glodt J, Valles-Colomer M, Koren O et al (2023) MetaPhlAn 4 profiling of unknown species-level genome bins improves the characterization of diet-associated microbiome changes in mice. Cell Rep 42(5):112464. https://doi.org/10.1016/j.celrep.2023.112464

Ordovas JM, Ferguson LR, Tai ES, Mathers JC (2018) Personalised nutrition and health. BMJ (Clinical Research Ed) 361:bmj.k2173. https://doi.org/10.1136/bmj.k2173

Qin J, Li R, Raes J, Arumugam M, Burgdorf KS, Manichanh C, Nielsen T, Pons N, Levenez F, Yamada T, Mende DR, Li J, Xu J, Li S, Li D, Cao J, Wang B, Liang H et al (2010) A human gut microbial gene catalogue established by metagenomic sequencing. Nature 464(7285):59–65. https://doi.org/10.1038/nature08821

Sempionatto JR, Montiel VR-V, Vargas E, Teymourian H, Wang J (2021) Wearable and mobile sensors for personalized nutrition. ACS Sensors 6(5):1745–1760. https://doi.org/10.1021/acssensors.1c00553

Sonnenburg JL, Bäckhed F (2016) Diet–microbiota interactions as moderators of human metabolism. Nature 535(7610):56–64. https://doi.org/10.1038/nature18846

Stance4health – Smart Technologies for personAlised Nutrition and Consumer Engagement (2018) Stance4health.com. https://www.stance4health.com/

The Integrative HMP (iHMP) Research Network Consortium, Proctor LM, Creasy HH, Fettweis JM, Lloyd-Price J, Mahurkar A, Zhou W, Buck GA, Snyder MP, Strauss JF III, Weinstock GM, White O, Huttenhower C (2019) The integrative human microbiome project. Nature 569(7758): 641–648. https://doi.org/10.1038/s41586-019-1238-8

Valdes AM, Walter J, Segal E, Spector TD (2018) Role of the gut microbiota in nutrition and health. BMJ (Clinical Research Ed) 361:k2179. https://doi.org/10.1136/bmj.k2179

Zeevi D, Korem T, Zmora N, Israeli D, Rothschild D, Weinberger A, Ben-Yacov O, Lador D, Avnit-Sagi T, Lotan-Pompan M, Suez J, Mahdi JA, Matot E, Malka G, Kosower N, Rein M, Zilberman-Schapira G, Dohnalová L, Pevsner-Fischer M et al (2015) Personalized nutrition by prediction of glycemic responses. Cell 163(5):1079–1094. https://doi.org/10.1016/j.cell.2015.11.001

The Role of Food Information Initiatives in Bringing Microbiome Awareness into the Diet

8

Carlos Abundancia, Laura Bosman, Kenia Fita Capdevila, Stephan Kampshoff, Camila Massri, Nina McGrath, Sofia Mendes, Florentin Ndizeye, Matteo Sabini, Debora Serra, Darya Silchenko, and Giacomo Sini

Abstract

Public interest in personalised nutrition and microbiome science has emerged in recent years, driven by advances in genomics, microbial research, and digital health tools. While these fields hold great potential for improving health outcomes, they remain scientifically evolving and prone to misinformation. Communicating these topics effectively is essential to bridging the gap between research and public understanding, ensuring that individuals receive accurate, evidence-based information to make informed dietary choices.

This chapter explores public awareness and misinformation surrounding the microbiome, presenting strategies for effective scientific communication and engagement. It highlights the role of food information initiatives, media campaigns, and interdisciplinary collaboration in promoting microbiome literacy. Additionally, it examines how healthcare professionals and researchers can be empowered to translate microbiome science into accessible guidance while managing public expectations. By addressing communication challenges, misinformation risks, and opportunities for engagement or collaborations, this chapter underscores the importance of transparent, evidence-based messaging in fostering a well-informed public capable of making sustainable, health-conscious decisions.

Keywords

Personalised nutrition · Human microbiome · Science communication · Misinformation · Awareness · Public engagement · Network

C. Abundancia · L. Bosman · K. F. Capdevila · S. Kampshoff (✉) · C. Massri · N. McGrath · S. Mendes · F. Ndizeye · M. Sabini · D. Serra · D. Silchenko · G. Sini
European Food Information Council (EUFIC), Brussels, Belgium
e-mail: stephan.kampshoff@eufic.org

M. P. Francino, J. Á. Rufián Henares (eds.), *The Microbiome and Personalized Nutrition*, The Microbiomes of Humans, Animals, Plants, and the Environment 6,
https://doi.org/10.1007/978-3-031-93902-0_8

8.1 Why Do We Need to Communicate About Personalised Nutrition and the Microbiome?

The general idea of personalised nutrition focuses on enhancing one's health and well-being through tailored dietary recommendations based on individual needs (Ordovas et al. 2018). The concept originated in the 1980s as a way to refine general dietary advice to specific patient needs (Crockett et al. 1988). However, it was not until the completion of the Human Genome Project and advancements in genomics and epigenomics in the mid-2000s that our modern understanding of personalised nutrition began to emerge and develop into a distinct scientific field (Holzapfel et al. 2022). Over the past two decades, researchers have integrated phenotypic, genotypic, metabolic, environmental, and lifestyle factors into personalised nutrition strategies.

At the same time, the field of human microbiome research, referring to the trillions of microorganisms residing in and on our bodies, has gained huge popularity amongst researchers and the general public. These microorganisms play a crucial role in various physiological processes, including digestion, immunity, and even mental health. Just like a fingerprint, each person's microbiome is unique. Recognising the microbiome's uniqueness and its influence on its host's metabolic status makes it a fundamental component of personalised nutrition (Sonnenburg and Bäckhed 2016).

The Challenge of Ambiguity in Public Communication Despite advances in understanding the human microbiome and individual nutritional needs, key concepts remain unclear and challenging to communicate. It is important to highlight both their significance for future public health applications and the existing knowledge gaps to the broader public.

For instance, this means recognising that both 'personalised nutrition' and the 'microbiome' are concepts still under scientific scrutiny and development. Scientists continue to debate the definition of 'personalised nutrition' and what it includes or excludes. As a result, it is often used interchangeably with other terms describing related concepts such as 'precision nutrition', 'targeted nutrition', 'nutrigenomics', 'nutrigenetics', or others (Ordovas et al. 2018; Mayes and Meloni 2024).

Similarly, the term 'microbiome' is still in the early stages of research. Research shows that each part of the body has a distinct microbiome (e.g. gut, lung, and skin), yet no consensus exists on what defines a 'healthy microbiome' (Joos et al. 2025). These ambiguities and the lack of clear definitions present significant challenges when communicating microbiome science to the public. Even as scientific understanding progresses, public communication must evolve accordingly. Establishing clear definitions and managing the uncertainty inherent in emerging fields is essential to maintaining scientific credibility and public trust.

Shaping Public Understanding and Managing Expectations Despite these uncertainties, effective communication about the microbiome and personalised nutrition remains crucial. Without clear, evidence-based communication, misinformation can flourish, leading to confusion, distrust, and misconceptions about diet and health. The widespread availability of health information on social media, blogs, and other platforms further complicates public understanding and risks reinforcing misleading claims (Shan et al. 2019; Schelkle and Galland 2020).

Addressing how to inform the public about topics with lacking scientific consensus, however, requires careful consideration. Effective communication strategies with clear and transparent messages help to bridge the gap between emerging scientific research and public knowledge. It involves explaining the current state of research, acknowledging the current gaps, and emphasising the importance of ongoing studies. By setting accurate expectations, food information initiatives can help the public make informed decisions without falling prey to misinformation or overly optimistic claims.

The Role of Science-Based Food Information Initiatives in Communicating Microbiome Science Science-based food information initiatives can play a pivotal role in bridging the gap between scientific research and public understanding. These initiatives encompass a range of activities, including public awareness campaigns, educational programmes, and media outreach, all designed to disseminate evidence-based information about nutrition and health. By making complex scientific concepts accessible and engaging, such initiatives help to demystify science and encourage informed lifestyle choices.

Through awareness events, webinars, workshops, and social media campaigns, recent research findings can be effectively communicated, enhancing public knowledge about the microbiome's role in health and disease. These efforts not only educate but also foster collaboration between academia, industry, policymakers, and the public, creating a shared space for dialogue and engagement.

The following chapter aims to provide an overview on how to improve public communication on the microbiome and personalised nutrition, ensuring that information remains scientifically accurate, accessible, and impactful. It explores public awareness and misinformation; tangible strategies for effective scientific communication; the role of campaigns, special events, and interdisciplinary collaboration in promoting microbiome awareness; and how researchers and healthcare professionals can be empowered to effectively engage with the public.

8.2 The Landscape of Public Awareness and Misinformation

In recent years, the field of 'gut health' and microbiome research has grown exponentially. This is driven by advancements in scientific research, media coverage, social media trends, the commercialisation of personalised nutrition products and services, and a growing public interest in health and wellness. The microbiome, which refers to the trillions of microorganisms living in and on our bodies, particularly in the gut, as well as their genes and metabolic activity has been recognised for its significant impact on health. While the term 'gut health' has gained widespread popularity in media, industry, and public discourse, it remains scientifically undefined and inconsistently used. However, while the heightened awareness has the potential to foster better health outcomes, it has also coincided with misinformation and unfounded claims which can undermine progress and public trust in microbiome research.

8.2.1 Rising Public Interest Driven by Scientific Advances and Commercialisation

Thousands of research papers on the gut microbiome are published every year (Walker and Hoyles 2023). This explosion of scientific research has been a significant driver of public interest. Studies linking the gut microbiome to various aspects of well-being—including immune function, mental health, and chronic disease prevention—have captured the public imagination and highlighted the potential for microbiome-focused interventions.

Personalised nutrition services, like ZOE (https://zoe.com), have boosted public interest in 'gut health'. By making gut microbiome testing and customised dietary advice more available, these companies use research to offer tailored health solutions, attracting customers who want proactive ways to manage their health. Other drivers include the release of popular science books, documentaries, and news articles bringing 'gut health' into the mainstream.

Digital metrics also highlight the growing public interest in the gut microbiome. For example, Google Trends data show a steady increase in searches related to 'gut health' and 'microbiome' over the past decade. Such searches include topics like 'gut health diets', 'cleanses', 'microbiome reboots', 'probiotic products', 'skincare routines', 'disease cures', and controversial treatments such as 'colonic hydrotherapy' and 'colonic reflorastation' (Marcon et al. 2021).

Social media platforms have also become hubs for health discourse. Hashtags such as #guthealth, #microbiome, and #healthygut boast millions of posts on Instagram and TikTok, where influencers and everyday users alike share tips, personal experiences, and product endorsements related to digestive wellness. The term 'gut health' is now being searched over 135,000 times monthly worldwide, the hashtag #guthealth has nearly 6 million posts on Instagram and over 725,000 posts on TikTok, and the most popular human gut microbiome YouTube videos cumulate over 62 million views (Chidambaram et al. 2022). However, as science is evolving, the rise in popularity also presents challenges in ensuring that the information disseminated is accurate, balanced, and evidence based.

8.2.2 The Rise of Misinformation and Its Impact on Public Health

Much of the online information on 'gut health' and the microbiome is inaccurate or misleading. The ease of content creation on the internet and social media means that anyone, regardless of their qualifications, can disseminate health advice. Furthermore, social media platforms enable users to customise their feeds by selecting preferred sources, which can create 'bubbles' or echo chambers where false information is repeatedly reinforced (Chidambaram et al. 2022). As a result, numerous unsupported or weakly supported claims have turned into 'facts' simply due to their repeated circulation (Walker and Hoyles 2023).

The Role of Popular Media in Spreading Misinformation The popular press has played a significant role in the spread of misinformation about the gut microbiome. Articles often discuss the microbiome in a non-research-specific context, focusing on generalised health benefits rather than specific research findings. Complex research is often oversimplified or greatly exaggerated, and concepts are reduced to rhetorical products that generate attention, attract readers, and promote related products (Marcon et al. 2021).

Popular press articles often portray the microbiome as a key factor in a wide range of health conditions, including digestive issues, obesity, cancer, allergies, skin health, and mental health conditions. They also typically outline steps to achieve the health benefits linked to the microbiome. Such actions commonly include food/drink intake (e.g. eating/drinking fermented foods, lentils, fresh fruits, and vegetables), taking probiotics, avoiding certain foods/drinks (e.g. artificial sweeteners, sugary foods/drinks, and caffeine), and avoiding antibiotics. However, very few articles offer critiques. Many fail to mention that the health benefits or current research on the microbiome may be unproven, uncertain, ineffective, or exaggerated. When critiques are present, they are often downplayed as simply reflecting preliminary research, which can give readers a false impression of how close these findings are to practical applications (Marcon et al. 2021).

Furthermore, many articles present the benefits of a 'healthy gut' as self-evident, often promoting dietary changes or probiotics without offering detailed evidence or context (Hooks et al. 2019). Finally, coverage tends to highlight observational studies, which are easier to communicate but less conclusive than clinical trials or systematic reviews that receive less attention (Prados-Bo and Casino 2021). As a result, the public often receives an incomplete picture of the science behind personalised nutrition and the gut microbiome.

Social Media and the Proliferation of Myths Social media platforms like TikTok, Instagram, and X (formerly known as Twitter) have become notorious for the dissemination of misleading and oversimplified health information (Suarez-Lledo and Alvarez-Galvez 2021; Diekman et al. 2023; Denniss et al. 2024). TikTok's most popular videos about 'gut health' often lack accurate advice, with influencers spreading inaccurate scientific claims and non-influencers promoting products and solutions based solely on anecdotal evidence (Frith et al. 2024; Lan and Tung 2024).

YouTube also hosts poor-quality and unreliable information about the gut microbiome (Chidambaram et al. 2022). Videos containing the most accurate information (i.e. videos produced by recognised institutions like universities and hospitals) often receive less engagement and likes, compared to content from independent creators, who often misrepresent or oversimplify the science. This underscores the concerning trend that low-quality, unreliable, and potentially misleading content is more easily accessed, accepted, and positively received by audiences who may not be familiar with the current scientific research on the gut microbiome. Furthermore, many creators fail to disclose financial interests in the products they promote.

Ethical Concerns and Potential Harm The rise of health misinformation raises significant ethical concerns. Misinformation can lead individuals to adopt practices that are not only ineffective but potentially detrimental to their health. For example, extreme elimination diets touted as solutions for gut healing may result in nutritional deficiencies, disordered eating patterns, and impaired gut function. Similarly, excessive and inappropriate use of supplements (e.g. high-dose probiotics or laxatives), detox programmes, colon cleanses, and other commercial products, may lack effectiveness and can pose risks for an individual's health (Sharp et al. 2009).

Furthermore, when promised results as advertised online fail to materialise, individuals may experience frustration, decreased trust in legitimate health advice, and reluctance to engage with evidence-based interventions in the future. This could delay proper treatment to patients or, on the other hand, lead to increased use of healthcare due to preventable complications or disease exacerbations. Healthcare providers may also face additional challenges in correcting misconceptions and guiding patients back to evidence-based practices, highlighting the broader systemic impacts of nutrition-related misinformation.

Particularly, the lack of clarity and transparency in benefit claims made by influencers and content creators is concerning. Such concerns are present in gut microbiome-based personalised nutrition services as well. Here, broad promises are made about the potential benefits, with disclaimers containing caveats that may not be prominent or clear enough to counteract the initial claims (Lee et al. 2023).

8.2.3 Public Awareness About the Microbiome and Personalised Nutrition

Consumer research indicates that while public interest in the human microbiome is rising, significant knowledge gaps persist (Rook and Zwart 2025). A global survey of 6500 individuals found that 64 % had heard of the term 'microbiome,' yet only 20 % could accurately define it (Prados 2023). Moreover, while many acknowledged its link to overall health, most struggled to articulate specific health effects or scientific evidence-based strategies for microbiome care (Williams et al. 2023a).

Demographic trends reveal notable disparities in microbiome awareness. Individuals with higher education levels and those employed in healthcare or life sciences consistently demonstrate greater knowledge, appreciation of its importance, and willingness to take actions such as dietary changes or participation in citizen initiatives (Rook and Zwart 2025). Conversely, older adults (especially those over 60) exhibit the lowest levels of understanding (Prados 2023).

Furthermore, the public's comprehension of microbiome-related health concepts remains limited. For example, only 28 % of respondents were familiar with the term 'dysbiosis' (an imbalance in the gut microbiome) and most lacked a clear understanding of its causes and consequences (Prados 2023). However, it is important to note that the term 'dysbiosis' should be used cautiously, as the concept of a 'healthy' microbiome is not clearly defined. Current research indicates that there is no clear understanding of what constitutes a 'healthy' or 'unhealthy' microbiome (Joos et al. 2025).

Additionally, many people, particularly adolescents, are unclear about how diet influences the gut microbiome. While many consumers recognise the link between the gut microbiome and overall health, they tend not to seek advice unless they experience specific symptoms (Williams et al. 2023a). Many college students express that while eating specifically for 'gut health' is important, it feels like 'too much trouble,' suggesting a need for targeted education on 'gut-healthy' eating practices for younger adults (Frith et al. 2024).

Consumer behaviour around the gut microbiome reflects a blend of accurate information and misconceptions. In the global survey of 6500 individuals, 57 % of surveyed individuals reported making lifestyle adjustments to support their microbiome, while they often lacked a precise understanding of its health implications (Prados 2023). Diet is commonly cited as the primary factor influencing ones gut microbiome, with fermented foods and probiotics being the most recognised interventions (Prados 2023). However, a prevalent misconception is that consuming probiotics or fermented foods automatically guarantees substantial health benefits, a belief that contradicts current scientific consensus and dietary guidelines.

Interest in Personalised Nutrition and Health Monitoring Despite knowledge gaps, there is a growing interest in personalised nutrition and monitoring microbiome health. Survey data reveal that 58–71 % of participants are willing to adjust their lifestyle, such as changing their diet or increasing exercise, to sustain or improve their microbiome health (Rook and Zwart 2025). Additionally, 40–54 % expressed interest in purchasing special products aimed at supporting their gut microbiome, and 49–57 % indicated a willingness to monitor their microbiome health through tools such as at-home testing kits or health apps (Rook and Zwart 2025). Parents, in particular, demonstrate greater adoption of practices to maintain a balanced microbiome compared to older adults (Prados 2023).

The Role of Healthcare Professionals and Trust in Information Sources Healthcare professionals, particularly family physicians, gastroenterologists, and dietitians, are trusted sources of information about the gut microbiome. About 44 % of people reported receiving guidance from healthcare professionals on maintaining a balanced microbiome, while 33 % had been informed about the disruptive effects of antibiotics on the gut microbiome (Prados 2023). However, despite the role of healthcare providers in disseminating this information, gaps in knowledge persist. For example, half of respondents lacked a clear understanding of what probiotics and prebiotics are, even though 62 % believed probiotics could help maintain microbiome health (Prados 2023).

8.2.4 Remaining Challenges

While digital communication platforms have enabled rapid and accessible dissemination of health information, they pose significant challenges regarding misinformation and clarity (Gatewood et al. 2020). Online sources, especially social media, remain key for spreading gut microbiome science but can also propagate inaccuracies. This paradox underscores the ongoing need to develop more reliable, evidence-based digital resources that meet the public's demand for clear, timely, and scientifically sound information on the gut microbiome.

Consumers frequently criticise current nutrition recommendations as outdated, which signals a potential for researchers and health professionals to update and expand accessible guidance (Williams et al. 2023a). Consumers are increasingly interested in understanding the science behind health recommendations and desire practical advice on integrating these insights into everyday routines. This feedback suggests that public health agencies and the scientific community need to collaborate in order to update recommendations and communicate them more clearly.

Complexity of the Human Microbiome One major challenge in microbiome science is that the microbiome's complexity makes it difficult to define a 'healthy' microbiome universally (Ma et al. 2018). While some dietary and lifestyle factors can influence the microbiome, determining the specific effects remains challenging. This ambiguity also impacts healthcare practitioners, as dietitians express concerns about their ability to provide microbiome-focused advice due to limited translatable research. Many report that their recommendations on the gut microbiome generally align with standard dietary guidelines, though further education will be essential as the evidence base grows (Williams et al. 2023b).

Addressing Public Misinformation Public scepticism of government health messaging presents another obstacle to effective consumer education. Using social media to engage consumers and explain evidence-based recommendations may help improve public adherence to health guidelines. Expanding on YouTube's COVID-19 misinformation policies, a similar approach could address gut health misinformation by regulating misleading content on digital platforms (Chidambaram et al. 2022).

Artificial intelligence (AI) could also play a role in filtering out misinformation, though current AI techniques struggle with the nuances of identifying misleading content about the microbiome. Ironically, AI is also used to create convincing fake content, further complicating efforts to control misinformation. Given these limitations, human oversight and reliable third-party verification remain critical to addressing misinformation about the gut microbiome (Aïmeur et al. 2023).

Food companies also share responsibility for consumer education, as clear packaging claims and practical guidance on product labels can help consumers make informed health decisions (Williams et al. 2023a). Many consumers value evidence-based advice but still prioritise personal preferences or anecdotal influences. This demonstrates a need for balanced communication that respects personal autonomy while conveying the evidence behind health recommendations.

Investing in education about the gut microbiome for the public, health professionals, and individuals is essential to improving overall health, beyond just digestive outcomes. Although most people recognise the importance of the gut microbiome, a comprehensive understanding of how diet, lifestyle, environmental factors, and mental health impact the microbiome is lacking. Broad educational efforts could help to convey that the gut microbiome is interconnected with these factors, emphasising that a healthy microbiome is not solely dependent on diet but also shaped by various lifestyle and environmental factors (Williams et al. 2023a).

Public awareness of personalised nutrition and the gut microbiome is growing, but misinformation remains a significant challenge. The widespread dissemination of misleading claims—particularly on social media and in commercialised health products—complicates consumer decision-making. Addressing these issues requires a collaborative approach between researchers, healthcare professionals, and science communicators to ensure that evidence-based, balanced information reaches the public.

8.3 Strategies for Effective Scientific Communication

The benefits of effective science communication can be multiple and varied. Some main reasons for communicating about the science of the microbiome and personalised nutrition include:

1. Increasing knowledge and awareness on the topic and aiding individuals to make more informed decisions about their health.
2. Promoting public engagement and input at all stages of the scientific process, thereby ensuring that research is more relevant to the public.
3. Increasing acceptance of new technologies and harnessing their potential.
4. Maintaining confidence in the scientific process and increasing scientific literacy.
5. Informing policy decision-making and generating funding support for further research in the field.

To achieve the desired impact, science communication must follow a tailored strategy that carefully considers the following aspects: the purpose of communication ('why'), the target audience(s) ('to whom'), the key messages ('what'), the communication channels and timing ('how'), the evaluation, and the engagement with relevant stakeholders. The following section outlines brief considerations for each of the key principles with a focus on communicating on the microbiome and its applications in personalised nutrition.

Purpose for Communicating Communication campaigns can have many different objectives. Before communicating, it is important to reflect on the end goal or purpose. Their aim might be, for example, to increase awareness or knowledge, or to persuade the target audience to take a specific action. For the communicator, understanding the desired outcome or impact will support the development of the overall strategy and maximise the chances of success. During this stage, performing a thorough evaluation of the social, political, or media environment can also be a valuable step for informing the development of an effective communication strategy. The surrounding environment affects how the public perceives communication around a specific topic, and this should influence the decisions you take when crafting your strategy.

Target Audience Communication is not a one-size-fits-all endeavour. All target audiences are different, and identifying who you are speaking to is crucial. Depending on who the communicator is aiming to reach, the values, perceptions, concerns, and information-seeking behaviour of your audience will be different. The better you know your audience, the better you can tailor your messages to their characteristics, concerns, feelings, and needs, which will ultimately lead to more successful communication.

There are many different tools available to gain a deeper understanding of a target audience, such as audience segmentation, building personas, focus groups, interviews, online deliberation research, observational research, surveys, experiments, and more. If you need to reach different target groups, each will require a tailor-made approach. For example, depending on sex or age, people's attitudes and intentions to adopt personalised nutrition programmes vary, with women and younger people tending to be more open. Communication initiatives will therefore need to be designed differently to reach target audiences that are less open, such as men or older demographics (Stewart-Knox et al. 2023).

Key Messages Once you have a good understanding of your communication purpose and your target audience, the next step is to develop a set of clear and concise key messages on which to base your communication efforts. This is the vital information you want your audience to understand and walk away with. The key messages will generally not be the final format of the communication effort, but simply a summary of the most important information to include in all further communications. When crafting key messages, consider the following principles:

- Tailor complexity to your audience. Adjust the depth of information based on the audience's level of scientific expertise.
- Use clear, relatable language. Avoid jargon and overly technical terms that may alienate non-experts.
- Connect with values and interests. Frame messages in a way that aligns with what matters most to your audience.
- Acknowledge uncertainties transparently. Clearly distinguish between established facts, ongoing research, and areas of uncertainty.

Communication Channels To ensure that your communications reach the desired audiences, they need to be delivered using the right channels. As part of your target audience analysis (step 2), you should have gained insight into where they usually receive information about nutrition and health. For example, while policymakers can be reached via professional conferences and events, reaching lay audiences will be more effective through traditional and social media, in schools, or via their healthcare providers.

The evolution of the traditional communication landscape over the last 30 years has introduced an interactive multi-directional conversation often incorporating audio-visual and rapid content sharing, rather than a one-way flow of information. Adapting your messages to channel-specific formats (e.g. using images, videos, hashtags, emojis, hyperlinks, and so on) is essential for more effective communication.

Timing The growing public interest in the microbiome presents an opportunity for increased public engagement by scientists. Aligning communication efforts with established awareness events, such as *World Microbiome Day* (June 27th) and International Microorganism Day (September 17th), can be a useful way to amplify messages at a time when public interest is high.

Evaluation and Optimisation An often-overlooked step in science communication is impact evaluation. To measure success, it is important to first set measurable and realistic goals, and plan which data you will collect and how you will evaluate success. Tools to measure impact and ensure that you are reaching your target audience can include surveys, feedback forms, website and social media analytics, and media monitoring software. The specific monitoring approach will depend on your communication channels and goals. Research and choose a tool that fits your needs in terms of cost, budget, resources, timeline, and ease of use.

By continuously monitoring and assessing the impact of communication, you can gather important feedback and learnings to improve future work.

Networks and Stakeholder Engagement Establishing a regular exchange with a diverse range of stakeholders working in the field of the microbiome and personalised nutrition might help to maximise the effectiveness and impact of your communication efforts. Facilitating dialogue among scientists, policymakers, industry representatives, NGOs, educators, and other communicators enhances the understanding of the topic from multiple perspectives. This stakeholder engagement has several benefits:

- It generates greater support and expands your reach through collaborative communication efforts shared across different networks and channels.
- It reduces the risk of conflicting or incoherent messages reaching the audience, leading to higher public trust.
- It encourages a more coordinated, cohesive approach to science communication in this space.

Challenges in Communicating About the Microbiome and Personalised Nutrition Science communication faces unique challenges when it comes to emerging research areas like the microbiome and personalised nutrition. While these fields hold great potential to enhance public health and well-being, their inherent complexity can make effective communication difficult. Some key challenges and potential strategies include:

- Addressing uncertainty: The microbiome and its applications in personalised nutrition are in their infancy, with much still unknown or uncertain. When communicating about scientific developments in the field, it is crucial to clearly explain uncertainties and the limitations of current research. This helps avoid setting unrealistic expectations and maintains public trust in the scientific process. Explain clearly what is known and what is not yet known or still under investigation.
- Engaging in two-way dialogue with the audience: The traditional deficit model of communication, characterised by the one-way flow of information from science to the public, is no longer sufficient. The rapid expansion of digital communication tools has provided a multitude of innovative ways to engage the public, but the number of options and resources needed to use them effectively can be challenging for individual communicators and organisations lacking sufficient resources. When getting started, it is important to take time to consider the right platform for your needs and skills, actively respond to questions and comments, participate in existing groups or communities, and track engagement metrics to understand which kinds of content perform well with your audience.
- Minimising scientific jargon or technical terminology: Using scientific terms or overly technical terminology can alienate non-scientific audiences, particularly on channels like social media. To avoid this, communicate in plain, everyday language, using storytelling and analogies to convey complex concepts and help your messages resonate with the audience. Incorporating visual content, such as images or infographics, can also help convey complex information in an impactful way (see Fig. 8.1).
- Online misinformation: Misrepresentation of science, oversimplified and sensationalist reporting on highly complex issues, and rapid online content sharing has resulted in conflicting and confusing information reaching the public. There is a staggering amount of information shared online regarding gut health and the microbiome, often from unqualified sources. Scientists play an important role, bringing in reliable scientific information into the public debate, while also supporting educational initiatives that promote public media literacy.

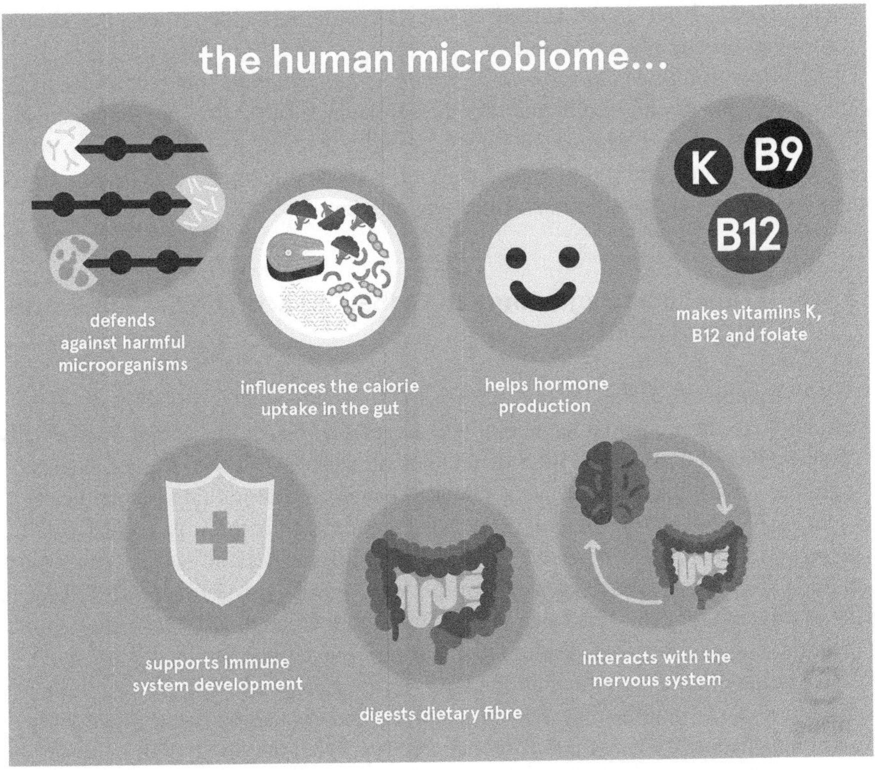

Fig. 8.1 Infographic about the human microbiome—the European Food Information Council (EUFIC) (www.eufic.org)

8.4 Engaging the Public Through Campaigns and Special Events

Public engagement is essential for fostering understanding and inspiring action on microbiome-related issues. Campaigns and public events serve as powerful tools to connect organisations with diverse audiences, and translate complex scientific concepts into accessible messages. By providing focused engagement and interactive opportunities, campaigns and events raise awareness while also fostering education, motivation, and long-term behavioural change. Well-coordinated public engagement efforts can enhance microbiome awareness and drive collective action.

One of the most effective ways to engage the public on a large scale is through global awareness days. These designated days bring critical issues to the forefront of public consciousness. Organisations such as the United Nations leverage these designated days, weeks, and even years or decades to highlight global challenges, fostering education, advocacy, and collective action. For example, the current decade (2021–2030) is the United Nations Decade of Healthy Ageing, promoting well-being among older populations.

Similarly, initiatives like *World Microbiome Day* leverage this global communication strategy to highlight microbiomes' role in health, food systems, and the environment. *World Microbiome Day* provides a dedicated platform to bridge scientific research and public understanding. It fosters dialogue, encourages action, and lays the groundwork for sustained engagement.

8.4.1 From Local Beginnings to Global Impact: The Case of World Microbiome Day

APC Microbiome Ireland launched *World Microbiome Day* in 2018 as a local initiative to raise awareness about the diverse world of microbiomes. Initially centred in Ireland, the event highlighted the critical roles of microbiomes in sustaining clean environments, mitigating climate change, and promoting human health.

In 2020, *World Microbiome Day* gained momentum through the EU-funded project *MicrobiomeSupport*, where APC Microbiome Ireland was a consortium partner. The event's reach across Europe was significantly broadened, thanks to the involvement of the European Food Information Council (EUFIC), the communication and dissemination leader for the *MicrobiomeSupport* project.

By collaborating with institutions, such as the European Food Safety Authority (EFSA) and microbiome research communities, and expanding engagement through multiple channels, *World Microbiome Day* evolved into a globally recognised initiative. This growth illustrates how local campaigns, with the right partnerships and resources, can evolve to achieve international visibility and impact. Collaboration with strategic partners not only extends a campaign's reach but also significantly enhances its credibility and engagement.

In today's digital age, campaigns are increasingly utilising interactive tools to engage audiences. *World Microbiome Day's* website includes quizzes, articles, competitions, and toolkits, designed to spark curiosity and facilitate learning. Interactive resources are particularly effective because they invite participants to actively explore the topic and allow the public to engage with the science behind microbiomes in a fun and approachable manner. Interactive educational materials foster a deeper understanding of the subject matter, moving beyond awareness-building to cultivating personal investment in the topic. The success of interactive content, such as quizzes or competitions, highlights the value of engaging the public in a manner that feels both accessible and informative.

In 2023, *World Microbiome Day* hosted its first-ever online celebratory event (Fig. 8.2), focused on the relationship between diets, microbiomes, and sustainable food systems. Featuring speakers from various EU-funded projects, the event attracted over 500 registrations, with 172 live attendees participating. The success of this event highlights the importance of virtual platforms in broadening the reach of public campaigns.

Fig. 8.2 Online event—World Microbiome Day 2023 (www.worldmicrobiomeday.com)

Online events enable global participation, breaking down geographical barriers and allowing experts, stakeholders, and the general public to come together in shared discussions. Such events also help bridge the gap between researchers and the public, offering an accessible platform where complex scientific ideas can be translated into everyday concepts.

8.4.2 Sustaining Public Engagement with Clear and Consistent Messaging

While special events and campaigns create significant short-term attention, their real power lies in building sustained public engagement via consistent informational and educational efforts. *World Microbiome Day* has fostered ongoing dialogue about microbiomes, evolving its message year after year. Similarly, EUFIC has continuously communicated the importance of microbiomes since 2013, when it participated in its first microbiome-related project, *MyNewGut*, consistently integrating this topic into broader public discourse through its own activities (e.g. social media campaigns, articles, and infographics).

Over recent years, EUFIC has observed how the microbiome has gained significant public attention. Microbes, once primarily associated with germs and food safety, are now increasingly recognised for their role in health. However, most public awareness still centres on the microbiome's impact on human health, as seen in data from online searches. There remains much work to be done in raising awareness about the microbiome's influence on other areas, such as food production and environmental sustainability.

Furthermore, there is ongoing confusion surrounding terms such as 'microbiome', 'microbiota', 'gut health', and 'gut flora', which are often used interchangeably. Educating the public on these distinctions is crucial. By clarifying these terms, people can better understand how their health is influenced by the interactions between microbes and their bodies. This empowers individuals to make informed decisions about their diet, lifestyle, and treatments. For instance, understanding that probiotics influence the microbiota, which in turn impacts the microbiome, helps set realistic expectations about health outcomes. Without these distinctions, there may be unrealistic or wrong expectations about immediate effects or health outcomes of such interventions.

Consistent engagement and clear communication are key to ensuring that public engagement translates into long-term behavioural change. By addressing misconceptions and continually educating the public, food information initiatives can cultivate a community of informed individuals who advocate for the cause well beyond the initial event.

8.4.3 Leveraging Social Media and Online Materials for Amplified Engagement

Social media plays a crucial role in amplifying public engagement initiatives. The *#MicrobiomeAmbassadors* campaign, initially launched as an activity of the *MicrobiomeSupport* project to give microbiome researchers and enthusiasts a voice, quickly expanded into a collaborative effort involving other EU-funded projects, such as *Human Microbiome Action*, *CIRCLES*, *DOMINO* or *Microbiomes4Soy*, and various organisations joining the cause.

Social media provides a direct channel for individuals and researchers to share insights. It enhances accessibility, fosters engagement, and allows experts to demystify complex microbiome concepts for the public. This connection allows experts to communicate directly with the public, making microbiome research more relatable. Social media's ability to give a voice to these stakeholders helps to increase the visibility of microbiome research and broadens public understanding of its significance.

Enhancing social media campaigns with online materials, such as articles and infographics, further strengthens outreach efforts. These materials are strategically optimised for search engines (SEO), ensuring they answer common questions that people search for online. By using SEO, food information initiatives like EUFIC ensure that their content reaches a wider audience, directly addressing the public's queries and concerns.

By analysing public search trends and data, organisations can ensure their content remains relevant, addressing evolving concerns and interests. Additionally, social media interactions generate valuable feedback, enabling organisations to refine messaging and tailor content to audience needs. This data-driven approach enables them to provide information that is both timely and useful.

Raising public awareness about microbiomes through campaigns and events is impactful, yet sustaining engagement requires more than momentary attention. Misinformation and knowledge gaps remain as challenges that can hinder meaningful public understanding. A strategic approach that combines scientific accuracy, clear messaging, and digital tools is essential. Collaboration between researchers, healthcare professionals, and science communicators ensures that evidence-based, accessible information reaches the public and drives lasting behavioural change.

8.5 Building and Leveraging Networks

Collaboration is essential for advancing microbiome research and its applications in food systems, public health, and personalised nutrition. Effective collaboration between researchers, healthcare professionals, citizens, policymakers, and industry stakeholders helps translate microbiome science into real-world applications. Networks create opportunities for knowledge exchange, interdisciplinary partnerships, and public engagement by providing a platform to facilitate knowledge sharing and cross-disciplinary connections.

Several EU-funded projects advance microbiome science by connecting researchers with relevant stakeholders. The following initiatives are only a selection of all those that have emerged over the years.

8.5.1 Sustainable Food Systems Network and Its Microbiome Subgroup

The *Sustainable Food Systems Network* (SFSN) is a multi-stakeholder platform that brings together over 2500 experts interested in and working towards the advancement of sustainable food systems. Its Microbiome Subgroup, launched in 2022, includes over 350 members and focuses on the role of microbiomes in food production, processing, and consumption.

The group was launched in 2022 with the *Microbiomes4Life* project cluster (including the EU-funded projects *MicrobiomeSupport*, *CIRCLES*, *HoloFood*, *MASTER*, and *SIMBA*), who co-created its goals and a shared code of conduct. The subgroup fosters collaboration by:

- *Breaking silos* between different disciplines and sectors
- *Providing access* to a diverse range of experts, research findings, and best practices
- *Enabling direct connections* through public discussions and private messaging within the platform

By integrating microbiome research within the broader food systems community, the subgroup ensures a holistic approach to sustainability and innovation.

8.5.2 Food 2030 Networks: Collaboration for Sustainable Food Systems

Several projects and initiatives are working on developing innovative solutions for a sustainable food system. However, these efforts often operate in isolation, targeting only a few components of the food system without the active engagement of all relevant stakeholders. The *FOOD 2030 Networks*, established under the *CLEVERFOOD* project, promote systemic food system transformation, supporting the EU's FOOD 2030 strategy. These networks offer a distinctive opportunity for EU-funded projects, Living Labs, partnerships, networks, and initiatives to unite and collaborate, breaking down silos and optimising the impact of their activities for long-term sustainability.

The *FOOD 2030 Networks* consist of two interconnected networks:

1. *FOOD 2030 Project Collaboration Network*
   A hub for projects and networks working towards sustainable, fair, and health-focused food systems aligning with EU policy priorities.
2. *FOOD 2030 Connected Lab Network*
   A community of Living Labs, policy councils, and co-creation initiatives supporting food system innovation at regional and national levels.

Through the collaboration fostered by the *FOOD 2030 Networks*, projects and initiatives can work on concrete actions, share successful practices, develop new competences, and influence policymaking and inform future EU legislation.

The systemic approach of the *FOOD 2030 Networks* is particularly relevant to the microbiome field. Microbiomes play a critical role in various aspects of the food system, from soil health and crop productivity to food safety and human health. By integrating microbiome science into FOOD 2030 initiatives, researchers can ensure that microbiome-related advancements contribute to sustainability and public health. The networks enable interdisciplinary projects to develop innovative solutions for food systems, facilitating connections between policymakers, researchers, and industry players to drive impactful change.

8.5.3 DRG4FOOD: Driving Digital Responsibility for Targeted Nutrition

The EU-funded *DRG4FOOD* project seeks for a data-driven food system that fosters trust throughout the food chain. It sees technology as a tool to drive innovation, rather than an end in itself. However, many technological developments are shaped by corporate and governmental needs, often neglecting user perspectives and exacerbating trust issues in food systems. Responsible technology development must prioritise privacy, transparency, and user control—especially when handling sensitive food data.

To support responsible innovation, *DRG4FOOD* launched two open calls (October 2023, May 2024), providing €1.9 million in funding to research groups, start-ups, and SMEs. This funding goes beyond financial support—it also facilitates training and mentoring, enabling recipients to exchange knowledge and integrate their solutions into a broader food technology ecosystem. The *Digital Responsibility Goals (DRGs)* serve as a unifying framework, guiding stakeholders towards shared principles of privacy, accountability, and social cohesion in digital food systems.

A core focus of *DRG4FOOD* is targeted nutrition, addressing the growing demand for personalised dietary advice. Through focus groups with more than 18 experts from the food and technology sectors, the project has developed a comprehensive definition of targeted nutrition. These discussions help ensure that scientific research is translated into practical, scalable applications, connecting academic insights with industry and policy decision-making.

> Delivering targeted diets and nutrition suggestions by applying a personalised approach towards more healthy diets, even considering personal variables. These variables encompass behavioural and emotional factors linked to food consumption, individual dietary preferences and restrictions, lifestyle, and so on (even in relation, e.g., due to personal beliefs, religious choices, weight management goals, environmental sustainability considerations, etc.)

Beyond research and industry collaboration, public engagement is a key pillar of *DRG4FOOD*. The project uses multiple communication approaches to connect with different stakeholders:

- *Mastodon and Data 4FoodCluster:* Engages digital ethics and data security professionals, fostering dialogue on responsible digital food technologies
- *Digital food podcast:* Brings together industry leaders, researchers, and the public to discuss digital responsibility and food innovation
- *Comic for teenagers:* Introduces younger audiences to key concepts in digital responsibility and food systems, helping build an informed future generation

Leveraging all these different networks is extremely important for a collective effort towards trustworthy, ethical, and transparent digital food systems and applications. Only via ongoing knowledge exchange and collaboration, responsible digital innovation becomes a long-term, integrated part of food system transformation.

8.5.4 DOMINO: Co-creating Fermented Food Products Through Living Labs

Emerging food technologies such as fermentation processes have the potential to contribute to sustainable, healthy, resilient, and circular food systems. However, there are various challenges to overcome for fermented food production to meet this potential, including gathering more in-depth knowledge about the production process, consumer acceptance of 'microbial foods', and the adoptability for producers to commercialise novel fermented food products.

The *DOMINO* project (2023–2028) uses Living Labs as a networking and co-creation tool to drive innovation in fermented foods. These real-world experimental environments bring together researchers, industry leaders, consumers, and policymakers to collaboratively shape food technologies that support human health, sustainability, and market adoption.

To ensure that fermentation technologies become a viable solution for sustainable, healthy, and resilient food systems, *DOMINO* Living Labs facilitate engagement between experts from various fields such as:

- *Science, and research & development:* Provide and develop in-depth knowledge of novel food production processes.
- *Industry and market:* Understand the commercial potential and market dynamics of new food products.
- *Society and consumers:* Provide insights into consumer perceptions, acceptance and behaviour towards new food products.
- *Producers:* Uptake emerging food technologies and commercialise products.

The *DOMINO* Living Labs enhance synergies by involving stakeholders from different disciplines and sectors, ensuring that a wide range of views and needs are considered, leading to more robust and accepted innovations. The collaborative nature of the Living Labs promotes co-creation, where stakeholders jointly develop food industry outputs, increasing the likelihood of successful adoption. Engaging stakeholders in the innovation process also helps build trust and disseminate knowledge.

8.5.5 Human Microbiome Action and the European Microbiome Centres Consortium

The *Human Microbiome Action* project (2021–2024) lays important groundwork for advancing microbiome research and innovation. The project focuses on paving the way for microbiome research and innovation by creating coherence and harmony in the way human microbiome research is and will be performed. The project not only aligns experts worldwide on key standards but also explored potential public health applications, shaping the future use of microbiome-based products in healthcare.

As a legacy of this effort, the *European Microbiome Centres Consortium* (EuroMCC) was established to drive forward developments in preventive nutrition and healthcare and act as a resource for European regulatory frameworks. The EuroMCC, founded by leading institutions across Europe, including INRAE (France), European Molecular Biology Laboratory (Germany), University College Cork (Ireland), University of Copenhagen (Denmark), and University of Trento (Italy), aims to promote and harmonise the development and validation of standards in microbiome research. By ensuring that these standards are consistent and universally recognised, the EuroMCC facilitates more accurate and reliable research outcomes.

The EuroMCC's main goals are to:

- Shape the human microbiome research and innovation community.
- Promote and harmonise standards development and validation.
- Advocate for the integration of scientific consensus into regulatory frameworks.
- Educate the public and healthcare professionals on microbiome research.

The consortium operates through six working groups focused on standardisation, analytical procedures, public health, scientific priorities, regulation, and communication.

8.5.6 CIRCLES: Engaging Citizens in the Microbiome of Food Chains

Microbes are everywhere in our living environment, yet how they contribute to the quality, productivity, safety, and environmental sustainability in the food system is largely unknown. *CIRCLES* (2018–2024) aims to understand the microbiome in the food system to optimise its application across six different food chains—spinach, tomatoes, poultry, swine, farmed and wild salmon, and farmed and wild seabream.

Engaging consumers to interactive learning and citizen participation is a cornerstone of *CIRCLES*' ambitions. By organising city tours and open-science events, the project creates opportunities for people to explore the invisible microbial world that surrounds them and understand its impact on food systems and human health.

One of *CIRCLES*' most dynamic public engagement efforts has been a series of city tours across Italy (Rimini, Bologna, and Mazara del Vallo), Finland (Helsinki), and Belgium (Brussels). These tours provide a hands-on experience, allowing citizens to:

- *Sample environmental microbiomes* in their homes and contribute to research.
- *Engage with scientists* in informal settings to ask questions and discuss findings.
- *Understand the role of microbiomes* in food production, sustainability, and safety.

Participants received sampling kits to test their household microbiome, helping to build the first city-wide map of house-floor microbes. Additionally, they had the opportunity to explore their own skin microbiome, observing microbial diversity through laboratory-prepared microbiological plates. Taking science out of the lab and into the real world strengthens the connection between the researchers and the public, making microbiome research tangible and relatable.

8.6 Empowering Researchers and Healthcare Professionals

Consumers are increasingly interested in research on personalised nutrition and the human microbiome. People, especially those with specific health conditions, seek to better understand their bodies and improve their well-being. However, public perceptions on the microbiome not necessarily align with academic perspectives. Traditional and social media sometimes misrepresent scientific studies, linking them to false claims or exaggerated conclusions. This overselling of science—without expert clarification—fuels misconceptions and unrealistic expectations (Shan et al. 2019).

Ultimately, this undermines scientific credibility and risks eroding public trust. In the field of personalised nutrition and microbiome research, this could slow innovation due to diminished societal support. Trust in science is essential for implementing new policies, processes, and technologies, particularly in health-related fields (Schelkle and Galland 2020).

Despite these challenges, the widespread interest in microbiome research presents an opportunity for researchers and healthcare professionals to engage and educate the public. Experts must bridge the knowledge gap by communicating current scientific understanding, including its limitations. Failing to do so allows misinformation to persist, harming both individuals and the integrity of science. By translating research into practical guidance, experts can enhance public health literacy and trust. Following are a few key strategies for empowering researchers and healthcare professionals to achieve these goals.

8.6.1 Continuous Education and Training

Staying up to date with microbiome research and personalised nutrition is increasingly difficult, even for experts. Misconceptions persist among healthcare professionals, highlighting the need for ongoing education. For instance, a recent survey found that nearly one-third of dietitians supported using commercially available gut microbiota test kits in their practice, despite ongoing debate and lack of standardised methods (Simon et al. 2023; Mitsou et al. 2024). Many healthcare providers also struggle to interpret microbiome test results, leaving them unprepared to advise patients (Rodriguez et al. 2024).

To address this challenge, continuous education and training must be prioritised. Effective training fosters dialogue between researchers and healthcare professionals, ensuring both groups understand research limitations, current and future applications, and patient needs. Because healthcare professionals act as a crucial gateway to the services available to patients, a shared understanding is critical for integrating microbiome-based applications into practice. Furthermore, these services should be validated and promoted by credible sources to build trust and ensure knowledge is effectively translated from research into practice (Beccia et al. 2023).

Educational initiatives must be tailored to specific audiences based on career stage and field of practice. Key approaches include:

- *Open-access publications* to lower barriers to knowledge exchange.
- *Conferences and workshops* for networking and in-depth discussions.
- *Academic curricula* integrating microbiome research into pre-graduate education.
- *Webinars and other e-learning opportunities* providing global access to training.

8.6.2 Transparent Communication

Public discourse on microbiome-based applications is highly active. However, the complexity of microbiome research allows misinformation to spread, leading to exaggerated claims and public distrust. To maintain credibility, researchers and healthcare professionals must apply clear and transparent communication strategies, as outlined earlier in this chapter (Schelkle and Galland 2020).

Several factors complicate communication in this field:

1. The microbiome's diverse effects on the host make it difficult to establish clear causal relationships.
2. A lack of standardisation in microbiome research results in uncertainty.
3. Research limitations and their impact on applications are often not sufficiently explained.

Effective communication requires transparency about these challenges. This fosters trust, promotes social equity, and supports informed policy discussions. Beyond general communication principles, managing public expectations about microbiome-based applications is essential. Clear, evidence-based messaging should be shared at multiple levels (Beccia et al. 2023):

- *Population-level outreach* through public institutions and health organisations.
- *Individual-level guidance* provided by healthcare professionals in direct patient interactions.

Public engagement should foster a two-way conversation rather than a top-down approach. Collaborating with experts from philosophy, behavioural science, and sociology can provide insight into public concerns and improve communication strategies (Schelkle and Galland 2020). Additionally, citizen science initiatives allow direct public involvement in research. Projects such as the *Flemish Gut Flora Project* and The *French Gut Project* include citizens by collecting stool samples, creating awareness and engagement. Similarly, Living Labs create collaborative environments for developing real-world solutions like microbiome-based food products or personalised nutrition services. By actively involving the public, scientists can empower individuals to take ownership of their health (Zwart 2024).

To improve accessibility, communication must be clear and relatable. Techniques such as storytelling, analogies, and familiar terminology make complex topics easier to understand (Schelkle and Galland 2020). As communication skills are often missing from academic and healthcare training, professionals should be equipped with tools to convey their messages effectively (Shan et al. 2019).

8.6.3 Interdisciplinary Collaboration

Philosopher Donna Haraway introduced the concept of sympoiesis, meaning 'making together' (Haraway 2016). This idea reflects the interconnected nature of life, including the relationship between humans and their microbiome. Just as the microbiome is embedded within its environment, the study and application of microbiome science should be approached through interdisciplinary collaboration (Zwart 2024).

Collaboration between experts across disciplines and international borders enhances knowledge exchange (Beccia et al. 2023). Partnering with professionals in various fields ensures balanced discussions on emerging research and public communication. For example, engaging with social scientists can reveal unmet needs and improve public health literacy strategies (Schelkle and Galland 2020).

Creating dedicated collaboration spaces is crucial. Initiatives such as the **European Microbiome Centres Consortium**, the **MicrobiomeSupport Association,** and the **Sustainable Food Systems Network** facilitate interdisciplinary engagement. These platforms bring together researchers, healthcare professionals, policymakers, and industry stakeholders to share knowledge and develop actionable solutions.

Empowering researchers and healthcare professionals is key to strengthening public health literacy and trust. Experts who stay informed, collaborate across disciplines, and communicate clearly can help shape public understanding of microbiome science. Providing professionals with the right tools and strategies fosters informed decision-making and empowers individuals to take a proactive role in managing their health and well-being. Rather than viewing the microbiome as something to control, citizens should be encouraged to act as stewards of their own health, embracing a balanced and informed approach to personalised nutrition (Zwart 2024).

8.7 Conclusion

Effective communication of microbiome science and personalised nutrition is essential in an era where public curiosity, misinformation, and evolving research intersect. While advancements in these fields have the potential to transform healthcare and dietary practices, their complexity presents significant communication challenges. Misinformation, particularly on digital platforms, continues to shape public perception, often leading to oversimplified or misleading health claims.

To counteract this, science-based food information initiatives, public engagement efforts, and interdisciplinary collaboration play a vital role in bridging the gap between research and public understanding. Strategies such as clear messaging, targeted education, and transparency in scientific uncertainties help to manage expectations and foster trust. Furthermore, equipping researchers and healthcare professionals with communication tools ensures that evidence-based insights reach broader audiences effectively.

By combining accurate information, public outreach, and responsible messaging, stakeholders across academia, industry, healthcare, and policy can create a well-informed society capable of making health-conscious decisions. Strengthening microbiome literacy will not only empower individuals but also enhance public confidence in nutrition science, contributing to a more evidence-driven and health-oriented future.

References

Aïmeur E, Amri S, Brassard G (2023) Fake news, disinformation and misinformation in social media: a review. Soc Netw Anal Min 13:30. https://doi.org/10.1007/s13278-023-01028-5

Beccia F, Causio FA, Hoxhaj I et al (2023) Integrating China in the international consortium for personalised medicine. A position paper on healthcare professionals' education and citizens' empowerment in personalised medicine. BMC Med Educ 23:438. https://doi.org/10.1186/s12909-023-04420-z

Chidambaram S, Maheswaran Y, Chan C et al (2022) Misinformation about the human gut microbiome in YouTube videos: cross-sectional study. JMIR Form Res 6:e37546. https://doi.org/10.2196/37546

Crockett SJ, Mullis RM, Perry CL (1988) Parent nutrition education: a conceptual model. J Sch Health 58:53–57. https://doi.org/10.1111/j.1746-1561.1988.tb05822.x

Denniss E, Lindberg R, Marchese LE, McNaughton SA (2024) #Fail: the quality and accuracy of nutrition-related information by influential Australian Instagram accounts. Int J Behav Nutr Phys Act 21:16. https://doi.org/10.1186/s12966-024-01565-y

Diekman C, Ryan CD, Oliver TL (2023) Misinformation and disinformation in food science and nutrition: impact on practice. J Nutr 153:3–9. https://doi.org/10.1016/j.tjnut.2022.10.001

Frith AL, Hitchen A, Hoffert A et al (2024) College student's experiences with the magic and myths of gut health. Curr Dev Nutr 8:102694. https://doi.org/10.1016/j.cdnut.2024.102694

Gatewood J, Monks SL, Singletary CR et al (2020) Social media in public health: strategies to distill, package, and disseminate public health research. J Public Health Manag Pract 26:489–492. https://doi.org/10.1097/PHH.0000000000001096

Haraway DJ (2016) Staying with the trouble: making kin in the Chthulucene. Duke University Press, Durham

Holzapfel C, Waldenberger M, Lorkowski S et al (2022) Genetics and epigenetics in personalized nutrition: evidence, expectations, and experiences. Mol Nutr Food Res 66:2200077. https://doi.org/10.1002/mnfr.202200077

Hooks KB, Konsman JP, O'Malley MA (2019) Microbiota-gut-brain research: a critical analysis. Behav Brain Sci 42:e60. https://doi.org/10.1017/S0140525X18002133

Joos R, Boucher K, Lavelle A et al (2025) Examining the healthy human microbiome concept. Nat Rev Microbiol 23:192–205. https://doi.org/10.1038/s41579-024-01107-0

Lan DH, Tung TM (2024) Exploring fake news awareness and trust in the age of social media among university student TikTok users. Cogent Soc Sci 10:2302216. https://doi.org/10.1080/23311886.2024.2302216

Lee K, Davies RG, Barnett J (2023) The presentation of gut microbiome-based personalized nutrition on the internet: simple and accessible, complex and inaccessible. Front Commun 8:974973. https://doi.org/10.3389/fcomm.2023.974973

Ma Y, Chen H, Lan C, Ren J (2018) Help, hope and hype: ethical considerations of human microbiome research and applications. Protein Cell 9:404–415. https://doi.org/10.1007/s13238-018-0537-4

Marcon AR, Turvey S, Caulfield T (2021) ‘Gut health’ and the microbiome in the popular press: a content analysis. BMJ Open 11:e052446. https://doi.org/10.1136/bmjopen-2021-052446

Mayes C, Meloni M (2024) Forgetting how we ate: personalised nutrition and the strategic uses of history. HPLS 46:14. https://doi.org/10.1007/s40656-024-00613-x

Mitsou EK, Katsagoni CN, Janiszewska K (2024) Knowledge of dietitians on gut microbiota in health—an online survey of the European Federation of the Associations of Dietitians (EFAD). Nutrients 16:621. https://doi.org/10.3390/nu16050621

Ordovas JM, Ferguson LR, Tai ES, Mathers JC (2018) Personalised nutrition and health. BMJ. https://doi.org/10.1136/bmj.k2173

Prados A (2023) How much do people know about the science of the microbiome? Gut Microbiota For Health. https://www.gutmicrobiotaforhealth.com/how-much-do-people-know-about-the-science-of-the-microbiome/

Prados-Bo A, Casino G (2021) Microbiome research in general and business newspapers: how many microbiome articles are published and which study designs make the news the most? PLoS One 16:e0249835. https://doi.org/10.1371/journal.pone.0249835

Rodriguez J, Cordaillat-Simmons M, Badalato N et al (2024) Microbiome testing in Europe: navigating analytical, ethical and regulatory challenges. Microbiome 12:258. https://doi.org/10.1186/s40168-024-01991-x

Rook O, Zwart H (2025) Awareness of human microbiome may promote healthier lifestyle and more positive environmental attitudes. Commun Med 5:39. https://doi.org/10.1038/s43856-025-00747-4

Schelkle B, Galland Q (2020) Microbiome research: open communication today, microbiome applications in the future. Microorganisms 8:1960. https://doi.org/10.3390/microorganisms8121960

Shan Y, Segre JA, Chang EB (2019) Responsible stewardship for communicating microbiome research to the press and public. Nat Med 25:872–874. https://doi.org/10.1038/s41591-019-0470-y

Sharp RR, Achkar J-P, Brinich MA, Farrell RM (2009) Helping patients make informed choices about probiotics: a need for research. Am J Gastroenterol 104:809–813. https://doi.org/10.1038/ajg.2008.68

Simon M, Sina C, Ferrario PG et al (2023) Gut microbiome analysis for personalized nutrition: the state of science. Mol Nutr Food Res 67:2200476. https://doi.org/10.1002/mnfr.202200476

Sonnenburg JL, Bäckhed F (2016) Diet–microbiota interactions as moderators of human metabolism. Nature 535:56–64. https://doi.org/10.1038/nature18846

Stewart-Knox BJ, Poínhos R, Fischer ARH et al (2023) Sex and age differences in attitudes and intention to adopt personalised nutrition in a UK sample. J Public Health (Berl) 31:1277–1283. https://doi.org/10.1007/s10389-021-01676-x

Suarez-Lledo V, Alvarez-Galvez J (2021) Prevalence of health misinformation on social media: systematic review. J Med Internet Res 23:e17187. https://doi.org/10.2196/17187

Walker AW, Hoyles L (2023) Human microbiome myths and misconceptions. Nat Microbiol 8: 1392–1396. https://doi.org/10.1038/s41564-023-01426-7

Williams GM, Tapsell LC, Beck EJ (2023a) Gut health, the microbiome and dietary choices: an exploration of consumer perspectives. Nutr Diet 80:85–94. https://doi.org/10.1111/1747-0080.12769

Williams GM, Tapsell LC, Beck EJ (2023b) Dietitians’ perspectives on the role of dietetics practice in ‘gut health’. Nutr Diet 80:95–103. https://doi.org/10.1111/1747-0080.12778

Zwart H (2024) Recommendations for responsible microbiome research and innovation. https://doi.org/10.5281/ZENODO.12604465

9783031939013